The Essentials of CAGD

The Essentials of
CAGD

GERALD E. FARIN
DIANNE HANSFORD

 **CRC Press**
Taylor & Francis Group
Boca Raton London New York

CRC Press is an imprint of the
Taylor & Francis Group, an **informa** business
AN A K PETERS BOOK

CRC Press
Taylor & Francis Group
6000 Broken Sound Parkway NW, Suite 300
Boca Raton, FL 33487-2742

© 2019 by Taylor & Francis Group, LLC
CRC Press is an imprint of Taylor & Francis Group, an Informa business

No claim to original U.S. Government works

ISBN 13: 978-0-367-45544-6 (pbk)
ISBN 13: 978-1-56881-123-9 (hbk)

Visit the Taylor & Francis Web site at
http://www.taylorandfrancis.com

and the CRC Press Web site at
http://www.crcpress.com

For Barbara & David

Contents

Preface   **xi**

1 The Bare Basics   **1**

    1.1 Points and Vectors 1
    1.2 Operations 3
    1.3 Products 5
    1.4 Affine Maps 7
    1.5 Triangles and Tetrahedra 8
    1.6 Exercises 11

2 Lines and Planes   **13**

    2.1 Linear Interpolation 13
    2.2 Line Forms 16
    2.3 Planes 18
    2.4 Linear Pieces: Polygons 19
    2.5 Linear Pieces: Triangulations 20
    2.6 Working with Triangulations 22
    2.7 Exercises 23

3 Cubic Bézier Curves   **25**

    3.1 Parametric Curves 25
    3.2 Cubic Bézier Curves 27
    3.3 Derivatives 30
    3.4 The de Casteljau Algorithm 32
    3.5 Subdivision 34
    3.6 Exploring the Properties of Bézier Curves 36
    3.7 The Matrix Form and Monomials 39
    3.8 Exercises 41

4 Bézier Curves: Cubic and Beyond   **43**

    4.1 Bézier Curves 43
    4.2 Derivatives Revisited 45
    4.3 The de Casteljau Algorithm Revisited 47

4.4   The Matrix Form and Monomials Revisited   48
4.5   Degree Elevation   49
4.6   Degree Reduction   52
4.7   Bézier Curves over General Intervals   54
4.8   Functional Bézier Curves   54
4.9   More on Bernstein Polynomials   55
4.10   Exercises   57

5   Putting Curves to Work   59

5.1   Cubic Interpolation   59
5.2   Interpolation Beyond Cubics   61
5.3   Aitken's Algorithm   63
5.4   Approximation   66
5.5   Finding the Right Parameters   68
5.6   Hermite Interpolation   68
5.7   Exercises   70

6   Bézier Patches   71

6.1   Parametric Surfaces   71
6.2   Bilinear Patches   72
6.3   Bézier Patches   75
6.4   Properties of Bézier Patches   77
6.5   Derivatives   78
6.6   Higher Order Derivatives   81
6.7   The de Casteljau Algorithm   83
6.8   Normals   83
6.9   Changing Degrees   87
6.10   Subdivision   88
6.11   Ruled Bézier Patches   90
6.12   Functional Bézier Patches   91
6.13   Monomial Patches   92
6.14   Exercises   93

7   Working with Polynomial Patches   95

7.1   Bicubic Interpolation   95
7.2   Interpolation using Higher Degrees   99
7.3   Coons Patches   100
7.4   Bicubic Hermite Interpolation   103
7.5   Trimmed Patches   105
7.6   Least Squares Approximation   106
7.7   Exercises   112

8   Shape   **115**

   8.1   The Frenet Frame 116
   8.2   Curvature and Torsion 118
   8.3   Surface Curvatures 122
   8.4   Reflection Lines 125
   8.5   Exercises 127

9   Composite Curves   **129**

   9.1   Piecewise Bézier Curves 129
   9.2 C^1 and G^1 Continuity 130
   9.3 C^2 and G^2 Continuity 132
   9.4   Working with Piecewise Bézier Curves 134
   9.5   Point-Normal Interpolation 135
   9.6   Exercises 136

10 B-Spline Curves   **139**

   10.1 Basic Definitions 139
   10.2 The de Boor Algorithm 143
   10.3 Practicalities of the de Boor Algorithm 149
   10.4 Properties of B-Spline Curves 150
   10.5 B-Splines: The Building Block 151
   10.6 Knot Insertion 156
   10.7 Periodic B-Spline Curves 159
   10.8 Derivatives 161
   10.9 Exercises 163

11 Working with B-Spline Curves   **165**

   11.1 Designing with B-Spline Curves 165
   11.2 Least Squares Approximation 166
   11.3 Shape Equations 169
   11.4 Cubic Spline Interpolation 171
   11.5 Cubic Spline Interpolation in a Nutshell 174
   11.6 Exercises 176

12 Composite Surfaces   **177**

   12.1 Composite Bézier Surfaces 177
   12.2 B-Spline Surfaces 181
   12.3 B-Spline Surface Approximation 183
   12.4 B-Spline Surface Interpolation 187
   12.5 Subdivision Surfaces: Doo-Sabin 188

12.6 Subdivision Surfaces: Catmull-Clark 190
12.7 Exercises 193

13 NURBS   195

13.1 Conics 195
13.2 Reparametrization and Classification 197
13.3 Derivatives 199
13.4 The Circle 199
13.5 Rational Bézier Curves 201
13.6 Rational B-Spline Curves 204
13.7 Rational Bézier and B-spline Surfaces 204
13.8 Surfaces of Revolution 205
13.9 Exercises 207

14 Hunting Geometry Bugs   209

Solutions to Selected Exercises   211

Notation   221

Bibliography   223

Index   225

Preface

The world of computing and communication has reached a level of visual content that was hard to imagine even ten years ago. Browsers download images, applets create animated sequences; entire movies are made with 100% computer-generated 3D imagery. Several of the underlying computational issues have their home in the field of *Computer Aided Geometric Design*, or CAGD. *The Essentials of CAGD* is an elementary introduction to those concepts which can be used to model the letters in this book as well as the "actors" in *Toy Story*.[1]

CAGD goes back to the 1950s when computers were used to drive numerically controlled milling machines in the automotive and aircraft industries. The basic tools that were developed then are *parametric curves and surfaces*. Now these are not only used in design and manufacturing (CAD/CAM) but also in computer graphics, computer animation, 3D visualization, reverse engineering, or robotics.

Several texts exist on the topic of CAGD. Why a new one? Some texts, for example Farin [9] or Hoschek/Lasser [15] assume a level of mathematical sophistication that is, in our experience, overwhelming for novices. Texts at lower math levels typically miss out on applications. The CAGD coverage in Computer Graphics texts is spotty. Hence we tried to create a comprehensive introduction that addresses a general audience and that covers many applications.

The Essentials of CAGD is intended for anyone who needs to learn the basic concepts of CAGD, be it as a first-time student or as a practitioner whose skills are a bit rusty. Its theoretical level is kept as low as possible—we usually substitute examples and images for exact proofs. *The Essentials of CAGD* is meant to be used at the freshman/sophomore undergraduate level. It serves as an introduction to CAGD for engineers or computer scientists. It is also an ideal companion text for a computer graphics class. Prerequsites for this text include basic computer graphics and linear algebra (such as provided in [10]).

---

[1] An animated movie produced by Pixar Studios.

The Essentials of CAGD approaches each topic from a geometric viewpoint. This is realized in three ways:

- **Sketches** illustrate the geometric elements of a concept.

- **Figures** illustrate a computer application of a concept.

- **Examples** illuminate algorithms by stepping the reader through a numerical application of a concept.

Exercises are listed at the end of each chapter. Solutions to selected exercises are given in Appendix A.

There is a short Bibliography, suggesting texts for reference and more advanced study of CAGD and related topics. The field of CAGD has its own journal—visit `http://www.journals.elsevier.com/computer-aided-geometric-design/`.

The Essentials of CAGD has a web site: `http://www.farinhansford.com/books/essentials-cagd/index.html`. This web site contains general information and updates. It also contains most PostScript files used in the book. See the web page for details on downloading these files. Also available are all data files referred to in the text. In the near future, the site will contain an errata page.

We like to thank the members of Arizona State University's PRISM project.[2] Particular thanks go to Mary Zhu for help with many graphics problems. Also thanks to M-S. Bae, J. McIntosh, A. Nasri, A. Razdan, H. Theisel. As usual, it was a pleasure to work with A K Peters during all stages of the publication process.

Gerald Farin   June 2000
Dianne Hansford   Paradise Valley, AZ

---

[2]For more information on this interdisciplinary project, visit `http://prism.asu.edu/`.

The Bare Basics

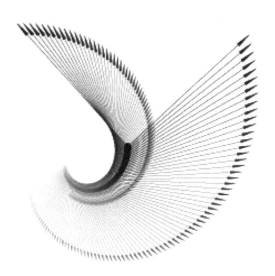

Figure 1.1.
A bare basic affine mapping of a vector.

In this first chapter, we cover some basic geometry. We also lay out the notation for the rest of the book. The reader should view this chapter as a reference that covers the prerequisites for the book.

1.1 Points and Vectors

Most geometry that we encounter happens in two or three dimensions. The first step toward formally defining geometry is to define a coordinate system. A *coordinate system* is defined by its origin $\mathbf{0}$ (a point) and two (or three) directions $\mathbf{e}_1, \mathbf{e}_2, \mathbf{e}_3$ (vectors). The system most commonly encountered for a 2D space:

$$\mathbf{0} = \begin{bmatrix} 0 \\ 0 \end{bmatrix}, \quad \mathbf{e}_1 = \begin{bmatrix} 1 \\ 0 \end{bmatrix}, \quad \mathbf{e}_2 = \begin{bmatrix} 0 \\ 1 \end{bmatrix}.$$

For a 3D space, things are not very different:

$$\mathbf{0} = \begin{bmatrix} 0 \\ 0 \\ 0 \end{bmatrix}, \quad \mathbf{e}_1 = \begin{bmatrix} 1 \\ 0 \\ 0 \end{bmatrix}, \quad \mathbf{e}_2 = \begin{bmatrix} 0 \\ 1 \\ 0 \end{bmatrix}, \quad \mathbf{e}_3 = \begin{bmatrix} 0 \\ 0 \\ 1 \end{bmatrix}.$$

The most basic geometric entity is the *point*, which simply denotes a 2D or 3D location. The collection of all 2D or 3D points is called 2D or 3D *affine space* or *Euclidean space*, and is denoted as $\mathbb{E}^2$ or $\mathbb{E}^3$.[1] If we pick a particular coordinate system for an affine space, then every point is defined by two or three coordinates.[2]

We denote 2D and 3D points by lower case boldface letters, such as $\mathbf{q}$ or $\mathbf{p}$. We write point coordinates as columns, for example

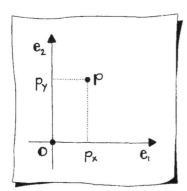

Sketch 1.
Coordinates of a 2D point.

$$\mathbf{q} = \begin{bmatrix} -1 \\ 2 \\ 4 \end{bmatrix} \quad \text{or} \quad \mathbf{p} = \begin{bmatrix} 1 \\ 2 \end{bmatrix}.$$

If we want to refer to a particular component, we denote it by q_x, q_y, q_z. In our example, we would have $p_x = 1$ and $p_y = 2$. Sketch 1 illustrates. In some situations, we will refer to the components of a point simply as x, y, z.

The difference of two 2D or two 3D points is called a 2D or 3D *vector*. The difference is simply computed coordinate by coordinate, for example

$$\begin{bmatrix} 2 \\ 2 \\ 0 \end{bmatrix} = \begin{bmatrix} 1 \\ 2 \\ 1 \end{bmatrix} - \begin{bmatrix} -1 \\ 0 \\ 1 \end{bmatrix},$$

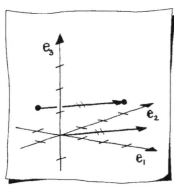

Sketch 2.
The difference of two points.

see Sketch 2.

We denote vectors with lower case boldface, just as we do points. Usually, there is no risk of confusion. A typical point-vector identity would then be written as $\mathbf{v} = \mathbf{p} - \mathbf{q}$, where $\mathbf{v}$ denotes the vector and $\mathbf{p}$ and $\mathbf{q}$ are the points. If we want to refer to a particular component, we denote it by v_x, v_y, v_z. Vectors "live" in a *linear space* or *real space*, which is denoted as $\mathbb{R}^2$ or $\mathbb{R}^3$.

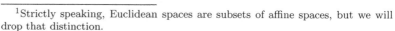

[1]Strictly speaking, Euclidean spaces are subsets of affine spaces, but we will drop that distinction.
[2]The same point will have different coordinates relative to a different coordinate system!

We have seen that affine and linear spaces are not the same. But since every affine space has an associated linear space (formed by the differences of all point pairs), it is common to plot points and vectors together, as in Sketch 3.

1.2   Operations

An essential operation for a point is a *translation*, which moves the point by a specified displacement. This "specified displacement" is typically defined by a vector. So if we translate a point $\mathbf{p}$ to another point $\hat{\mathbf{p}}$ by a vector $\mathbf{v}$, then we have

$$\hat{\mathbf{p}} = \mathbf{p} + \mathbf{v}.$$

Translations change point coordinates, but they have no effect on vector coordinates.[3] Consider two points $\mathbf{a}$ and $\mathbf{b}$. A difference vector is $\mathbf{w} = \mathbf{a} - \mathbf{b}$. If we translate both points by a vector $\mathbf{v}$, then the translated points are given by $\mathbf{a} + \mathbf{v}$ and $\mathbf{b} + \mathbf{v}$. If we now take the difference, we see that it is again given by $\mathbf{a} - \mathbf{b}$. Sketch 4 illustrates.

We may combine vectors by forming *linear combinations*, such as[4]

$$\mathbf{v} = \mathbf{v}_1 + \mathbf{v}_2.$$

If we interpret these vectors as representing translations, then the translation given by $\mathbf{v}$ is simply the translation given by $\mathbf{v}_1$, followed by one given by $\mathbf{v}_2$. Thus we may also write

$$\hat{\mathbf{p}} = \mathbf{p} + \mathbf{v}_1 + \mathbf{v}_2,$$

denoting that $\hat{\mathbf{p}}$ is obtained from $\mathbf{p}$ by applying two translations.

More generally, a linear combination of vectors is written as

$$\mathbf{v} = \alpha_1 \mathbf{v}_1 + \alpha_2 \mathbf{v}_2 + \ldots + \alpha_n \mathbf{v}_n,$$

with real numbers $\alpha_1, \ldots, \alpha_n$ and vectors $\mathbf{v}_1, \ldots, \mathbf{v}_n$.

While vectors may be combined using any real numbers as factors in a linear combination, that is not true of points: If we combine two points $\mathbf{p}$ and $\mathbf{q}$ to yield a third point $\mathbf{x}$ by setting

$$\mathbf{x} = \alpha \mathbf{p} + \beta \mathbf{q}, \tag{1.1}$$

---

[3]We asssume that the vector is defined as the difference of two points.

[4]We multiply a vector by a scalar by multiplying each of the vector's components by that scalar.

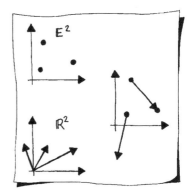

Sketch 3.
Affine and linear spaces illustrated separately and together.

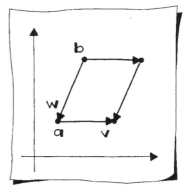

Sketch 4.
Vectors are unaffected by translations.

then it is mandatory that $\alpha + \beta = 1$.[5] Such special linear combinations are called *barycentric combinations*. Otherwise, the relation (1.1) would not be true after we apply a translation to all involved points.

EXAMPLE 1.1

Let

$$\mathbf{p} = \begin{bmatrix} 0 \\ 1 \end{bmatrix} \quad \text{and} \quad \mathbf{q} = \begin{bmatrix} 1 \\ 1 \end{bmatrix}.$$

Form the combination $\mathbf{x} = 2\mathbf{p} + \mathbf{q}$:

$$\mathbf{x} = \begin{bmatrix} 1 \\ 3 \end{bmatrix}.$$

Now translate both $\mathbf{p}$ and $\mathbf{q}$ by a vector $\mathbf{v} = \begin{bmatrix} 2 \\ 2 \end{bmatrix}$, resulting in

$$\hat{\mathbf{p}} = \begin{bmatrix} 2 \\ 3 \end{bmatrix} \quad \text{and} \quad \hat{\mathbf{q}} = \begin{bmatrix} 3 \\ 3 \end{bmatrix}.$$

Now form $\hat{\mathbf{x}} = 2\hat{\mathbf{p}} + \hat{\mathbf{q}}$:

$$\hat{\mathbf{x}} = \begin{bmatrix} 7 \\ 9 \end{bmatrix}.$$

This is not the same as applying our translation to $\mathbf{x}$! Sketch 5 illustrates.

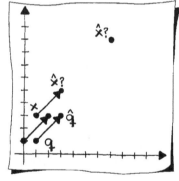

Sketch 5.
An illegal point operation.

The next example shows that the legal operation of forming barycentric combinations will remain valid after translations.

EXAMPLE 1.2

Let

$$\mathbf{p} = \begin{bmatrix} 0 \\ 1 \end{bmatrix} \quad \text{and} \quad \mathbf{q} = \begin{bmatrix} 1 \\ 1 \end{bmatrix}.$$

Form the combination $\mathbf{x} = 0.5\mathbf{p} + 0.5\mathbf{q}$:

$$\mathbf{x} = \begin{bmatrix} 0.5 \\ 1 \end{bmatrix}.$$

---

[5]For the special case $\alpha = \beta = 0.5$, we call $\mathbf{x}$ the *midpoint* of $\mathbf{p}$ and $\mathbf{q}$.

Now translate both $\mathbf{p}$ and $\mathbf{q}$ by a vector $\mathbf{v} = \begin{bmatrix} 2 \\ 2 \end{bmatrix}$, resulting in

$$\hat{\mathbf{p}} = \begin{bmatrix} 2 \\ 3 \end{bmatrix} \quad \text{and} \quad \hat{\mathbf{q}} = \begin{bmatrix} 3 \\ 3 \end{bmatrix}.$$

Now form $\hat{\mathbf{x}} = 0.5\hat{\mathbf{p}} + 0.5\hat{\mathbf{q}}$:

$$\hat{\mathbf{x}} = \begin{bmatrix} 2.5 \\ 3 \end{bmatrix}.$$

This is the same as applying our translation to $\mathbf{x}$! Sketch 6 illustrates.

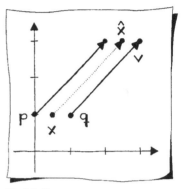

Sketch 6.
A legal point operation.

A barycentric combination of points

$$\mathbf{p} = \alpha_1 \mathbf{p}_1 + \ldots + \alpha_n \mathbf{p}_n$$

needs $\alpha_1 + \ldots + \alpha_n = 1$. A common term used to describe this condition is to say that the α_i form a *partition of unity*.

Another definition: When we form a barycentric combination $\mathbf{x} = a\mathbf{p} + b\mathbf{q}$, then we say that the *ratio* of the three points is given by

$$\text{ratio}(\mathbf{p}, \mathbf{x}, \mathbf{q}) = b : a = \frac{b}{a}. \tag{1.2}$$

See Sketch 7 for examples.

In that sketch, the ratio pair $b : a$ is written so that b is on the segment $\overline{\mathbf{px}}$, and a is on the segment $\overline{\mathbf{xq}}$. This tip is helpful for remembering how to write the "algebraic" relationship of (1.1).

For the configuration of Sketch 7, we also observe

$$\text{ratio}(\mathbf{p}, \mathbf{x}, \mathbf{q}) = \frac{\|\mathbf{x} - \mathbf{p}\|}{\|\mathbf{q} - \mathbf{x}\|}$$

with $\|\mathbf{x} - \mathbf{p}\|$ denoting the length of a vector; see Section 1.3. For configurations with $\mathbf{x}$ not between $\mathbf{p}$ and $\mathbf{q}$, we have to use signed lengths.

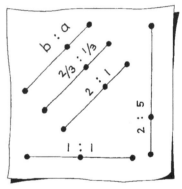

Sketch 7.
The point $\mathbf{x}$ is in the ratio $b : a$ with respect to $\mathbf{p}$ and $\mathbf{q}$, respectively.

1.3   Products

Two more operations for vectors will be used frequently: dot and cross products. If $\mathbf{v}$ and $\mathbf{w}$ are two vectors, their *dot product* is defined by

$$\mathbf{v}\mathbf{w} = v_x w_x + v_y w_y$$

for 2D vectors and by

$$\mathbf{v}\mathbf{w} = v_x w_x + v_y w_y + v_z w_z$$

for 3D vectors. The dot product is a scalar, related to the angle α between the vectors $\mathbf{v}$ and $\mathbf{w}$:

$$\cos(\alpha) = \frac{\mathbf{v}\mathbf{w}}{\|\mathbf{v}\|\|\mathbf{w}\|},$$

where $\|\mathbf{v}\|$ denotes the *length* of a vector:

$$\|\mathbf{v}\| = \sqrt{\mathbf{v}\mathbf{v}}.$$

In particular, the dot product of two vectors is zero when they are *perpendicular* to each other: In that case, $\cos(\alpha) = 0$ and therefore, $\alpha = 90°$.

Dot products are *symmetric*, meaning that

$$\mathbf{v}\mathbf{w} = \mathbf{w}\mathbf{v}.$$

The *cross product*, also known as the *vector product*, is used mostly for 3D vectors. It is the vector defined by

$$\mathbf{v} \wedge \mathbf{w} = \left[\begin{array}{c} v_y w_z - v_z w_y \\ v_z w_x - v_x w_z \\ v_x w_y - v_y w_x \end{array} \right].$$

The cross product of two vectors is perpendicular to both of them. Its length is given by

$$\|\mathbf{v} \wedge \mathbf{w}\| = \|\mathbf{v}\|\|\mathbf{w}\| \sin(\alpha).$$

This length is equal to the area of the parallelogram spanned by $\mathbf{v}$ and $\mathbf{w}$. Sketch 8 illustrates. As an application, we may use the vector product to calculate the area of a triangle. See Section 1.5

The cross product is the zero vector if $\alpha = 0$, i.e., if $\mathbf{v}$ and $\mathbf{w}$ are multiples of each other. In that case, $\mathbf{v}$ and $\mathbf{w}$ are called *linearly dependent*.

Cross products are *antisymmetric*, meaning that

$$\mathbf{v} \wedge \mathbf{w} = -\mathbf{w} \wedge \mathbf{v}.$$

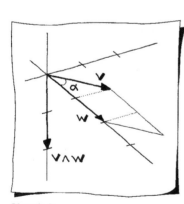

Sketch 8.
The cross product.

1.4   Affine Maps

We may translate geometric figures, rotate them around a point or an axis, or simply scale them. In all cases we apply *affine maps*. Affine maps map points to points, lines to lines, and planes to planes. [6]

If $\mathbf{p}$ is a point (i.e., $\mathbf{p} \in I\!\!E^2$), then an affine map results in a point $\hat{\mathbf{p}} \in I\!\!E^2$, defined as follows:

$$\hat{\mathbf{p}} = A\mathbf{p} + \mathbf{v}. \qquad (1.3)$$

Here, A is a 2×2 matrix and $\mathbf{v} \in I\!\!R^2$. If $\mathbf{p}$ were a 3D point, then A would be a 3×3 matrix and $\mathbf{v} \in I\!\!R^3$.

One important property of affine maps is that they leave the ratio of three collinear points unchanged. This is a key characterization that will be used many times throughout this text.

EXAMPLE 1.3

Let three collinear 2D points be given by

$$\begin{bmatrix} 0 \\ -1 \end{bmatrix}, \quad \begin{bmatrix} 1 \\ 0 \end{bmatrix}, \quad \begin{bmatrix} 2 \\ 1 \end{bmatrix},$$

and let an affine map be given by

$$\hat{\mathbf{x}} = \begin{bmatrix} 1 & 1 \\ 0 & 1 \end{bmatrix} \mathbf{x} + \begin{bmatrix} 0 \\ 1 \end{bmatrix}.$$

The images of our three points are

$$\begin{bmatrix} -1 \\ 0 \end{bmatrix}, \quad \begin{bmatrix} 1 \\ 1 \end{bmatrix}, \quad \begin{bmatrix} 3 \\ 2 \end{bmatrix}.$$

In both cases, the middle point is the midpoint of the two other points! See Sketch 9 for an illustration.

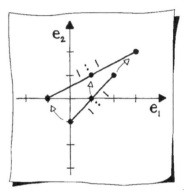

Sketch 9.
Affine maps preserve midpoints.

Another important property of affine maps is that they take parallel lines to parallel lines. If two lines are parallel before an affine map is applied to them, then they will be parallel after the mapping. Conversely, if the lines do not intersect before the mapping, they will not intersect after the mapping. The same applies to planes.

---

[6]Lines and planes are defined in Chapter 2.

An affine map is composed of two parts: the matrix A and the translation vector $\mathbf{v}$. The matrix A represents a *linear map*. Technically, the linear map should be applied to vectors only[7]. However, to simplify notation, we apply A to a point $\mathbf{x}$ rather than the vector $\mathbf{x} - \mathbf{0}$. Linear maps are capable of scalings, rotations, reflections, shears, projections, or combinations of all. Some examples:

$$\text{scale}: \begin{bmatrix} 2 & 0 \\ 0 & 2 \end{bmatrix} \qquad \text{reflection}: \begin{bmatrix} 1 & 0 \\ 0 & -1 \end{bmatrix}$$

$$\text{rotation}: \begin{bmatrix} \cos(\alpha) & -\sin(\alpha) \\ \sin(\alpha) & \cos(\alpha) \end{bmatrix} \qquad \text{shear}: \begin{bmatrix} 1 & 3 \\ 1 & 1 \end{bmatrix}$$

$$\text{projection}: \begin{bmatrix} 1 & 0 \\ 0 & 0 \end{bmatrix}$$

Try applying these matrices to the $\mathbf{e}_1$ and $\mathbf{e}_2$ vectors and see what happens!

1.5   Triangles and Tetrahedra

We all know triangles from high school, so the quick review here is mostly for establishing our notation. A 2D triangle T is formed by three noncollinear points $\mathbf{a}$, $\mathbf{b}$, $\mathbf{c}$.

Its area, $\text{area}(\mathbf{a}, \mathbf{b}, \mathbf{c})$, is easily computed using a 3×3 determinant:

$$\text{area}(\mathbf{a}, \mathbf{b}, \mathbf{c}) = \frac{1}{2} \begin{vmatrix} 1 & 1 & 1 \\ \mathbf{a} & \mathbf{b} & \mathbf{c} \end{vmatrix}, \tag{1.4}$$

which is short for

$$\text{area}(\mathbf{a}, \mathbf{b}, \mathbf{c}) = \frac{1}{2} \begin{vmatrix} 1 & 1 & 1 \\ a_x & b_x & c_x \\ a_y & b_y & c_y \end{vmatrix}.$$

Now let $\mathbf{p}$ be an arbitrary point inside T. Our aim is to write it as a combination of the triangle vertices, in a form like this:

$$\mathbf{p} = u\mathbf{a} + v\mathbf{b} + w\mathbf{c}. \tag{1.5}$$

We know one thing already: The right hand side of this equation is a combination of points, and so the coefficients must sum to one:

$$u + v + w = 1.$$

---

[7]Vectors are elements of a linear space.

Otherwise, we would not have a barycentric combination!

We next observe that (1.5) is short for two equations, one for each coordinate. Together with $u+v+w=1$, we then have three equations in the three unknowns u,v,w. Their solution is easily obtained using Cramer's rule:

$$u = \frac{\text{area}(\mathbf{p},\mathbf{b},\mathbf{c})}{\text{area}(\mathbf{a},\mathbf{b},\mathbf{c})}, \tag{1.6}$$

$$v = \frac{\text{area}(\mathbf{p},\mathbf{c},\mathbf{a})}{\text{area}(\mathbf{a},\mathbf{b},\mathbf{c})}, \tag{1.7}$$

$$w = \frac{\text{area}(\mathbf{p},\mathbf{a},\mathbf{b})}{\text{area}(\mathbf{a},\mathbf{b},\mathbf{c})}, \tag{1.8}$$

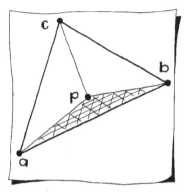

We call (u,v,w) *barycentric coordinates* and denote them by bold-face: $\mathbf{u} = (u,v,w)$. Although they are not independent of each other (we may set $w = 1 - u - v$), they behave much like "normal" coordinates: If $\mathbf{p}$ is given, then we can find $\mathbf{u}$ from (1.6)-(1.8).[8] If $\mathbf{u}$ is given, then we can find $\mathbf{p}$ from (1.5). It is, by the way, not necessary that $\mathbf{p}$ be inside T. If it is not, then some of u,v,w will be negative—their sum still being unity, of course. This implies that the area calculation must produce a *signed* area. Sketch 10 illustrates that the barycentric coordinates are ratios of areas; the coordinate w is the quotient of the shaded area and the entire triangle area.

The three vertices of the triangle have barycentric coordinates

Sketch 10.
Barycentric coordinates as the ratios of areas.

$$\mathbf{a} \cong (1,0,0),$$
$$\mathbf{b} \cong (0,1,0),$$
$$\mathbf{c} \cong (0,0,1).$$

The $\cong$ symbol will be used to indicate the barycentric coordinates of a point.

A triangle may also be defined in 3D: then it is given by three 3D points $\mathbf{a},\mathbf{b},\mathbf{c}$. Its area is given by

$$\text{area}(\mathbf{a},\mathbf{b},\mathbf{c}) = \frac{1}{2}\|[\mathbf{b}-\mathbf{a}] \wedge [\mathbf{c}-\mathbf{a}]\|.$$

---

[8]Notice that there is no need to *compute* w from (1.8), rather compute $w = 1 - u - v$.

EXAMPLE 1.4

Let

$$\mathbf{a} = \begin{bmatrix} 0 \\ 0 \\ -1 \end{bmatrix}, \quad \mathbf{b} = \begin{bmatrix} 0 \\ 2 \\ 0 \end{bmatrix}, \quad \mathbf{c} = \begin{bmatrix} 1 \\ 0 \\ 3 \end{bmatrix}.$$

Then

$$\text{area}(\mathbf{a}, \mathbf{b}, \mathbf{c}) = \frac{\sqrt{69}}{2}.$$

This area is one-half the length of the vector $\begin{bmatrix} 8 \\ 1 \\ -2 \end{bmatrix}$.

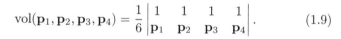

In 3D, we may also define a *tetrahedron*. It is given by four 3D points $\mathbf{p}_1, \mathbf{p}_2, \mathbf{p}_3, \mathbf{p}_4$. If these points do not all lie in one plane, then the volume of the tetrahedron is given by

$$\text{vol}(\mathbf{p}_1, \mathbf{p}_2, \mathbf{p}_3, \mathbf{p}_4) = \frac{1}{6} \begin{vmatrix} 1 & 1 & 1 & 1 \\ \mathbf{p}_1 & \mathbf{p}_2 & \mathbf{p}_3 & \mathbf{p}_4 \end{vmatrix}. \tag{1.9}$$

EXAMPLE 1.5

The four points

$$\mathbf{p}_1 = \begin{bmatrix} 0 \\ 0 \\ 0 \end{bmatrix}, \quad \mathbf{p}_2 = \begin{bmatrix} 2 \\ 0 \\ 0 \end{bmatrix}, \quad \mathbf{p}_3 = \begin{bmatrix} 3 \\ 3 \\ 0 \end{bmatrix}, \quad \mathbf{p}_4 = \begin{bmatrix} 1 \\ 1 \\ 2 \end{bmatrix}$$

define a tetrahedron—see Sketch 11—with volume

$$\frac{1}{6} \begin{vmatrix} 1 & 1 & 1 & 1 \\ 0 & 2 & 3 & 1 \\ 0 & 0 & 3 & 1 \\ 0 & 0 & 0 & 2 \end{vmatrix} = \frac{12}{6} = 2.$$

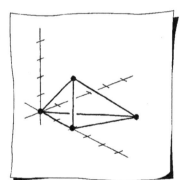

Sketch 11.
A tetrahedron.

1.6   Exercises

1. Given four points $\mathbf{p}, \mathbf{q}, \mathbf{r}, \mathbf{y} \in E^3$, is the point $\mathbf{x}$ formed as

$$\mathbf{x} = 3\mathbf{p} + 2\mathbf{q} - 5\mathbf{r} + \mathbf{y}$$

   a barycentric combination? Give the reason for your answer.

2. Adding or subtracting two vectors yields another vector. Create your own sketch to illustrate the *parallelogram rule*; the sum and difference of two vectors are the diagonals of the parallelogram defined by the vectors.

3. Let three points be given by

$$\mathbf{p}_1 = \begin{bmatrix} 0 \\ 0 \end{bmatrix}, \quad \mathbf{p}_2 = \begin{bmatrix} 1 \\ 2 \end{bmatrix}, \quad \mathbf{p}_3 = \begin{bmatrix} 4 \\ 8 \end{bmatrix}.$$

   Compute ratio$(\mathbf{p}_1, \mathbf{p}_2, \mathbf{p}_3)$.

4. Then repeat with

$$\mathbf{p}_1 = \begin{bmatrix} -1 \\ 2 \end{bmatrix}, \quad \mathbf{p}_2 = \begin{bmatrix} 0 \\ 1 \end{bmatrix}, \quad \mathbf{p}_3 = \begin{bmatrix} 1 \\ 0 \end{bmatrix}.$$

5. Form the triangle $\mathbf{a}, \mathbf{b}, \mathbf{c}$ with the three 2D points

$$\mathbf{a} = \begin{bmatrix} -1 \\ -1 \end{bmatrix}, \quad \mathbf{b} = \begin{bmatrix} 1 \\ -1 \end{bmatrix}, \quad \mathbf{c} = \begin{bmatrix} 0 \\ 2 \end{bmatrix}.$$

   Compute the barycentric coordinates of the points

$$\mathbf{p}_1 = \begin{bmatrix} 0 \\ 0 \end{bmatrix}, \quad \mathbf{p}_2 = \begin{bmatrix} 0 \\ -1 \end{bmatrix}, \quad \mathbf{p}_3 = \begin{bmatrix} 0 \\ 2 \end{bmatrix}.$$

6. Consider the three 3D points

$$\mathbf{a} = \begin{bmatrix} -1 \\ 0 \\ 1 \end{bmatrix}, \quad \mathbf{b} = \begin{bmatrix} 1 \\ 2 \\ -2 \end{bmatrix}, \quad \mathbf{c} = \begin{bmatrix} 1 \\ -1 \\ 1 \end{bmatrix}.$$

   What is the area of the triangle formed by them?

7. Repeat for the points

$$\mathbf{a} = \begin{bmatrix} -1 \\ 2 \\ 1 \end{bmatrix}, \quad \mathbf{b} = \begin{bmatrix} 1 \\ 2 \\ -2 \end{bmatrix}, \quad \mathbf{c} = \begin{bmatrix} 1 \\ 2 \\ 1 \end{bmatrix}.$$

8. Barycentric coordinates with respect to a triangle in 2D are formed as ratios of areas. In a similar manner, describe barycentric coordinates with respect to a tetrahedron in 3D.

Lines and Planes 2

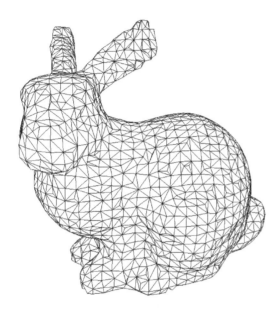

Figure 2.1.
A triangulation of the "Stanford bunny."

Next, we cover the basics of lines and planes. Linear interpolation is one of the most fundamental operations, both geometrically and computationally. After all, the only arithmetic operations a computer can perform exactly are addition and multiplication—the components of linear interpolation. The fundamentals of lines and planes and their various representations are part of this chapter. Finally, the geometric entities, polygons and triangles, are explored.

2.1 Linear Interpolation

Two points (assume 2D for now) $\mathbf{p}$ and $\mathbf{q}$, define a straight line, or line for short. How can we describe all points on this line?

One way is to imagine a particle, represented by a point $\mathbf{x}$, traversing the line, starting at $\mathbf{p}$ at time $t = 0$, passing through $\mathbf{q}$ at time $t = 1$, and continuing on. We also assume that the speed of our particle is constant. Since the location of $\mathbf{x}$ depends on the time t, we also write $\mathbf{x}(t)$ instead of just $\mathbf{x}$.

Where is $\mathbf{x}(t)$ at any given time t? Since we know $\mathbf{x}(0) = \mathbf{p}$ and $\mathbf{x}(1) = \mathbf{q}$, it seems reasonable to expect

$$\mathbf{x}(\frac{1}{2}) = \frac{1}{2}\mathbf{p} + \frac{1}{2}\mathbf{q}.$$

These three instances follow the general pattern

$$\mathbf{x}(t) = (1 - t)\mathbf{p} + t\mathbf{q}, \tag{2.1}$$

which is the *parametric form* of a line. This is also called *linear interpolation*.

EXAMPLE 2.1

Let

$$\mathbf{p} = \begin{bmatrix} 20 \\ 2 \end{bmatrix}, \quad \mathbf{q} = \begin{bmatrix} -10 \\ -10 \end{bmatrix}.$$

At $t = \frac{1}{3}$, we have

$$\mathbf{x}(\frac{1}{3}) = \frac{2}{3}\mathbf{p} + \frac{1}{3}\mathbf{q} = \begin{bmatrix} 10 \\ -2 \end{bmatrix},$$

which is illustrated in Sketch 12.

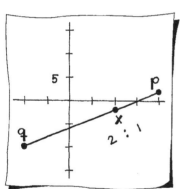

Sketch 12.
Linear interpolation in 2D.

Thus, for any real number t, we can compute $\mathbf{x}(t)$. The notion of time is intuitive, but not really necessary. We will typically refer to t as a *parameter*. Its *domain* is the set of real numbers, and (2.1) provides a map of the real line to the line spanned by $\mathbf{p}$ and $\mathbf{q}$: It maps a parameter t in the domain of the map to a point $\mathbf{x}(t)$ in the *range* of the map. We also say that t is the *preimage* of $\mathbf{x}(t)$.

This map is in fact an *affine map*. Why? Consider the ratio of the three points $\mathbf{p}, \mathbf{x}(t), \mathbf{q}$. From the definition of a ratio (1.2), it follows that

$$\text{ratio}(\mathbf{p}, \mathbf{x}(t), \mathbf{q}) = \frac{t}{1 - t}.$$

Next we check the ratio of the three corresponding preimages. Since $t = (1 - t) \cdot 0 + t \cdot 1$, it follows that

$$\text{ratio}(0, t, 1) = \frac{t}{1 - t}.$$

These two ratios are the same—our map is indeed affine.

To describe this affine map in the context of (1.3) we need to *embed* the 1D points $0, t, 1$ in 2D. We can do this by simply associating them with the points

$$\begin{bmatrix} 0 \\ 0 \end{bmatrix}, \quad \begin{bmatrix} t \\ 0 \end{bmatrix}, \quad \begin{bmatrix} 1 \\ 0 \end{bmatrix}.$$

Then the exact scale, rotation, and translation needed to map $[0, 1]$ onto $[\mathbf{p}, \mathbf{q}]$ may be determined. With A and $\mathbf{v}$ defined,

$$\mathbf{x} = A \begin{bmatrix} t \\ 0 \end{bmatrix} + \mathbf{v}$$

This is certainly more work than (2.1), and thus is not recommended!

The parametric form easily extends to 3D. If $\mathbf{p}$ and $\mathbf{q}$ are 3D points, then a 3D point $\mathbf{x}(t)$ on the straight line through them is again given by (2.1).

As stated earlier, the parameter t can be any real number. In other words, it is not restricted to be in the interval $[0, 1]$. A 3D example for this:

EXAMPLE 2.2

Let

$$\mathbf{p} = \begin{bmatrix} 1 \\ 1 \\ -1 \end{bmatrix}, \quad \mathbf{q} = \begin{bmatrix} -1 \\ 1 \\ 2 \end{bmatrix}.$$

At $t = -1$, we have

$$\mathbf{x}(-1) = 2\mathbf{p} - \mathbf{q} = \begin{bmatrix} 3 \\ 1 \\ -4 \end{bmatrix},$$

which is illustrated in Sketch 13.

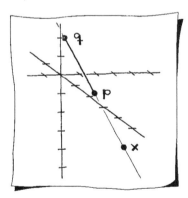

Sketch 13.
Linear interpolation in 3D.

A straight line is infinite; in many practical applications, one only needs the finite part between $\mathbf{p}$ and $\mathbf{q}$. This is referred to as a *line segment*.

Particularly in the context of the time analogy for the parametric line, we might want the line segment to be associated with an interval other than $[0, 1]$. Suppose instead, we want the line segment to correspond to $[a, b]$. Let u be the parameter value associated with $[a, b]$. To apply (2.1): construct the affine map which takes $u \in [a, b]$ to $t \in [0, 1]$, or

$$t = \frac{u - a}{b - a}. \tag{2.2}$$

This makes

$$1 - t = \frac{b - u}{b - a}.$$

The parameter u is referred to as the *global parameter* and t is the *local parameter*. The process of changing intervals is called a *parameter transformation*.

EXAMPLE 2.3

Let

$$\mathbf{p} = \left[\begin{array}{c} 1900 \\ 1K \end{array} \right], \quad \mathbf{q} = \left[\begin{array}{c} 2000 \\ 100K \end{array} \right].$$

Suppose $\mathbf{p}$ is associated with time 1900 (the year) and $\mathbf{q}$ with 2000. What is the data point for 1990?

The year 1990 is a global parameter; the local parameter for the line through $\mathbf{p}$ and $\mathbf{q}$ corresponds to

$$t = \frac{1990 - 1900}{2000 - 1900} = \frac{9}{10}.$$

The data point for 1990 is

$$\mathbf{x}(\frac{9}{10}) = \frac{1}{10}\mathbf{p} + \frac{9}{10}\mathbf{q} = \left[\begin{array}{c} 1990 \\ 90100 \end{array} \right].$$

2.2   Line Forms

Equation (2.1) is often referred to as the *parametric form* of a straight line. Most people first encounter straight line equations in the form

$y = ax + b$, also called *explicit*. The parametric form is more general: While the straight line through

$$\begin{bmatrix} 0 \\ 0 \end{bmatrix}, \begin{bmatrix} 0 \\ 1 \end{bmatrix}$$

is easily represented in parametric form, it cannot be expressed as $y = ax + b$. Also, the parametric form has a natural 3D extension while the explicit one does not.

If a line is given in the explicit form, how can we retrieve the parametric form? A moment's reflection should reveal that there is no "the" parametric form: As soon as we can compute any two points on the line, we can write down the corresponding parametric form. Two obvious candidates are the points corresponding to $x = 0$ and $x = 1$, giving

$$\mathbf{x}(t) = \begin{bmatrix} x(t) \\ y(t) \end{bmatrix} = (1 - t) \begin{bmatrix} 0 \\ b \end{bmatrix} + t \begin{bmatrix} 1 \\ a + b \end{bmatrix}.$$

We may rewrite (2.1) as

$$\mathbf{x}(t) = \mathbf{p} + t(\mathbf{q} - \mathbf{p}).$$

The vector $\mathbf{v} = \mathbf{q} - \mathbf{p}$ indicates the direction of the line. If $\mathbf{v}^\perp$ is perpendicular[1] to $\mathbf{v}$, then for any $\mathbf{x}$ on the line,

$$\mathbf{v}^\perp[\mathbf{x} - \mathbf{p}] = 0. \tag{2.3}$$

For the moment, use x and y as the coordinates of $\mathbf{x}$, then the equation is of the form

$$ax + by = c \tag{2.4}$$

and is called the *implicit form* or the *point–normal form*. Any point with coordinates x and y satisfying (2.4) is on the line.

EXAMPLE 2.4

Let an implicit line be given by

$$3x - 2y = 1.$$

---

[1]For instance, $\begin{bmatrix} -v_y \\ v_x \end{bmatrix}$ is perpendicular to $\mathbf{v}$.

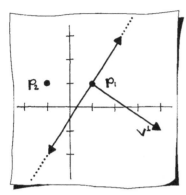

Sketch 14.
Points on and off a line.

Of the two points

$$\mathbf{p}_1 = \begin{bmatrix} 1 \\ 1 \end{bmatrix}, \quad \mathbf{p}_2 = \begin{bmatrix} -1 \\ 1 \end{bmatrix},$$

$\mathbf{p}_1$ is on the line, while $\mathbf{p}_2$ is not. See Sketch 14. To sketch the implicit line, first find one point on the line—which we have with $\mathbf{p}_1$. The vector $\mathbf{v}^\perp = \begin{bmatrix} 3 \\ -2 \end{bmatrix}$, which makes

$$\mathbf{v} = \begin{bmatrix} 2 \\ 3 \end{bmatrix} \quad \text{or} \quad \mathbf{v} = \begin{bmatrix} -2 \\ 3 \end{bmatrix}.$$

2.3 Planes

The 3D analog to a 2D line is a *plane*. A plane is defined by three noncollinear points $\mathbf{p}, \mathbf{q}, \mathbf{r}$. Any point $\mathbf{x}$ of the form

$$\mathbf{x} = u\mathbf{p} + v\mathbf{q} + w\mathbf{r} \quad \text{with} \quad u + v + w = 1 \tag{2.5}$$

is in that plane. To see why this is true, recall (1.9) for the volume of a tetrahedron. Applying it to our four points, we have

$$\text{vol}(\mathbf{p}, \mathbf{q}, \mathbf{r}, \mathbf{x}) = \frac{1}{6} \begin{vmatrix} 1 & 1 & 1 & 1 \\ \mathbf{p} & \mathbf{q} & \mathbf{r} & \mathbf{x} \end{vmatrix}$$

which vanishes since the last column, as per (2.5), is a linear combination of the first three columns. A tetrahedron with zero volume is a plane, and so $\mathbf{x}$ is indeed in the plane spanned by $\mathbf{p}, \mathbf{q}, \mathbf{r}$.

Rewriting (2.5), we have

$$\mathbf{x} = \mathbf{r} + u(\mathbf{p} - \mathbf{r}) + v(\mathbf{q} - \mathbf{r}). \tag{2.6}$$

This is the *point–vector form* of a plane, still in parametric form.

The *implicit form* is obtained as follows: Since $\mathbf{n} = [\mathbf{p} - \mathbf{r}] \wedge [\mathbf{q} - \mathbf{r}]$ is orthogonal, or *normal* to the plane, any point $\mathbf{x}$ in the plane must satisfy

$$\mathbf{n}[\mathbf{x} - \mathbf{r}] = 0, \tag{2.7}$$

which, after some algebra, becomes

$$ax + by + cz = d, \qquad (2.8)$$

where the coordinates of $\mathbf{x}$ are referred to as x, y, z. Equation (2.7) is known as the *point–normal form* of the plane.

The implicit form offers a quick test to see if a point $\mathbf{x}$ lies in the plane or not: Simply insert its coordinates into the implicit form. If they satisfy the equation, $\mathbf{x}$ is in the plane; otherwise, it is not.

EXAMPLE 2.5

Given the three points

$$\mathbf{p} = \begin{bmatrix} 1 \\ 0 \\ 1 \end{bmatrix}, \quad \mathbf{q} = \begin{bmatrix} 0 \\ 1 \\ 1 \end{bmatrix}, \quad \mathbf{r} = \begin{bmatrix} 1 \\ 1 \\ 0 \end{bmatrix},$$

find the implicit equation of the plane through them.

First, we compute a normal vector $\mathbf{n}$ to the plane:

$$\mathbf{n} = [\mathbf{p} - \mathbf{r}] \wedge [\mathbf{q} - \mathbf{r}] = \begin{bmatrix} -1 \\ -1 \\ -1 \end{bmatrix}.$$

The implicit form is then given by

$$\begin{bmatrix} -1 \\ -1 \\ -1 \end{bmatrix} \cdot \begin{bmatrix} x - 1 \\ y - 1 \\ z \end{bmatrix} = 0$$

or

$$x + y + z = 2.$$

Sketch 15 illustrates.

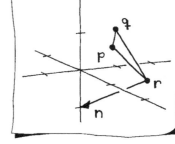

Sketch 15.
A plane through three points.

2.4   Linear Pieces: Polygons

In many applications, objects are formed by piecing together linear or planar pieces.

In the 2D case, line segments are pieced together to form *polygons*. A polygon is given by a set of points $\mathbf{p}_1, \ldots, \mathbf{p}_N$, where each $\mathbf{p}_i$ is

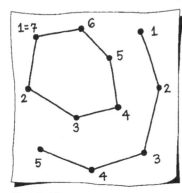

Sketch 16.
Open and closed polygons

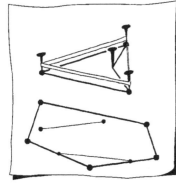

Sketch 17.
Convexity tests for polygons.

connected to $\mathbf{p}_{i+1}$ by a straight line segment. The points are called *vertices* and the line segments are called *edges*.

Polygons may be *open* or *closed*. Referring to Sketch 16, we see that this is mostly a matter of labeling the vertices: In the open case, $\mathbf{p}_1$ and $\mathbf{p}_N$ are distinct, whereas in the closed case, they are identical.

Closed polygons are often classified as to whether they are *convex* or nonconvex. The "rubber band" test is a good conceptual test for determining the convexity. If nails were put in the paper at each vertex, and then a rubber band was fit around the nails, then the polygon is convex if the rubber band touches each nail. The area enclosed by the rubber band is known as the *convex hull*. Another test: connecting any two points in or on the edges of the polygon forms a line segment. All points on this line segment must be inside the polygon. This must be true for any pair of points selected. Sketch 17 illustrates.

2.5   Linear Pieces: Triangulations

Triangles are the most fundamental entity in computer graphics; most rendering boils down to determining how a triangular facet interacts with the lighting model. CAD/CAM has historically used triangles as a centerpiece geometric entity for computations such as tool paths and finite element analysis (FEM). The reason: Computations are very simple and fast.[2]

More recently, triangulations are generated through the use of *laser digitizers*. Those instruments scan an object using laser rays, just like the laser devices used at the checkout in a supermarket. The scanning generates x, y, z coordinates of points on the digitized object. Built-in software then creates a triangulation. The resulting triangulations tend to be large, typically tens or hundred thousands of points. For an interesting example, see the "Digital Michelangelo" web site `http://www-graphics.stanford.edu/projects/mich/`. It describes a data set with two billion triangles. Figure 2.1 illustrates a triangulated 3D object.

Now for the actual definition: A set of connected triangles in 2D or in 3D is called a *triangulation* of a set of points. A triangulation must satisfy the following conditions:

1. The vertices of the triangles consist of the given points.

2. The interiors of any two triangles do not intersect.

---

[2]Recall the important coordinate system for triangles, the *barycentric coordinates*, discussed in Section 1.5.

3. If two triangles are not disjoint, then they share a vertex or have a coinciding edge.

4. All triangles are oriented consistently such that their normals point "outward."

Less stringent definitions might be acceptable for some applications. There are many possibilities for a data structure to define a triangulation. A commonly used structure includes:

- A *point list* containing the coordinates of the vertices.

- A *triangle list* containing triples of pointers into the point list, where each triple indicates the vertices of a triangle.

- A *neighbor list* containing triples of pointers into the triangle list, where the ith triple indicates the triangles neighboring the ith triangle. Also, the jth entry in a triple corresponds to the neighbor opposite the jth vertex. If there is no neighbor, then this is marked by an entry -1.

Depending on the application at hand, one should decide on the optimal data structure: It could be in the form of arrays or of doubly linked lists.

EXAMPLE 2.6

For the triangulation of Sketch 18, the point list is given by

$$\mathbf{p}_0, \mathbf{p}_1, \mathbf{p}_2, \mathbf{p}_3, \mathbf{p}_4, \mathbf{p}_5, \mathbf{p}_6.$$

In a "real" example, each $\mathbf{p}_i$ would be replaced by its coordinates.
The triangle and neighbor lists:

triangle	vertices	neighbors
0	1, 6, 4	5, −1, 2
1	5, 0, 3,	−1, 3, −1
2	1, 2, 6	4, 0, −1
3	5, 3, 6	5, 4, 1
4	2, 5, 6	3, 2, −1
5	4, 6, 3	3, −1, 0

This has to be read as follows: Triangle 0 consists of points $\mathbf{p}_1, \mathbf{p}_6, \mathbf{p}_4$. This triangle has two neighbors. Opposite $\mathbf{p}_1$ is Triangle 5. Opposite $\mathbf{p}_6$, there is no neighbor; this is marked by a -1 entry. Opposite $\mathbf{p}_4$ is Triangle 2. The other entries in the list follow the same pattern.

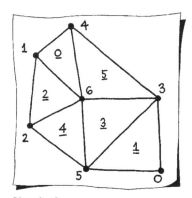

Sketch 18.
A triangulation data structure.

The data structure used in STL (Stereo Lithography Language) format is an example of a bad data structure. In this format, the data points are listed multiply—each triangle is given explicitly by its vertices. Additionally, no neighbor information is given. These two features make processing such triangulations difficult.

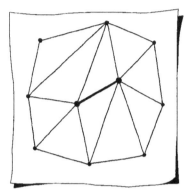

Sketch 19.
A triangulation with a highlighted edge.

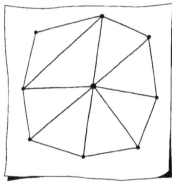

Sketch 20.
A triangulation after edge collapse.

2.6 Working with Triangulations

In many cases, triangulations are far too dense for an efficient representation of an object. What is called for then is a reduction in size, also known as *decimation* The goal is to remove as many triangles as possible, while still staying as close to the initial triangulation as possible—and, of course, while keeping a valid triangulation.

The basic idea is this: if the triangles in Sketch 19 are (within a tolerance) part of a plane, then they may be replaced by fewer triangles which would still describe the initial shape adequately.

One way to reduce the number of triangles is to *collapse an edge*. In Sketch 19 we marked an edge that we would like to collapse; in fact, we will reduce it to its midpoint. The resulting triangulation is shown in Sketch 20. As you see, we have reduced the number of triangles while still maintaining a valid triangulation. This process of edge collapses is repeated until it cannot be applied any more without unacceptable changes to the geometry.

A question that remains is: when is the geometry around an edge flat enough so that it can be collapsed safely? Testing this is done by a *flatness test*. There are several possibilities for such a test; we discuss a plane-based one here.

First, we define the *star* $\mathbf{p}^\star$ of a point $\mathbf{p}$. It is the set of all triangles having $\mathbf{p}$ as a vertex. In Sketch 18, $\mathbf{p}_6^\star$ consists of all triangles formed by it and points $\mathbf{p}_1, \ldots, \mathbf{p}_5$.

Let $\mathbf{p}$ and $\mathbf{q}$ form an edge in the triangulation; we want to know if we can collapse it. We need to check if all triangles formed by $\mathbf{p}^\star \cup \mathbf{q}^\star$ are sufficiently planar. For this, we first construct a plane: it is defined by a point and a normal. The point is the centroid of all points forming $\mathbf{p}^\star \cup \mathbf{q}^\star$, and the normal is the average normal of all triangles in $\mathbf{p}^\star \cup \mathbf{q}^\star$. If all points are within a given tolerance to this plane, then the edge formed by $\mathbf{p}$ and $\mathbf{q}$ may be collapsed. Figure 2.2 shows the triangulation of Figure 2.1 after decimation was applied to it.

As we coarsen a triangulation by collapsing edges, we represent the underlying object at several levels of resolution: very fine initially

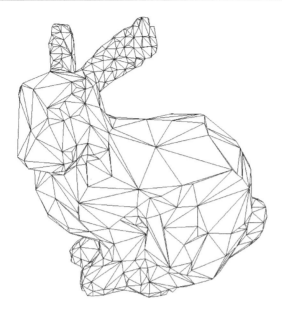

Figure 2.2.
A decimated Stanford bunny.

and reasonably coarse at the end. Being able to represent geometry at various levels of detail is referred to as *multiresolution*. An obvious application is geometry transmission through the internet. With proper bookkeeping, it is possible to transmit a triangulation at its coarsest level first, then step by step adding more detail as transmission progresses.

2.7   Exercises

1. Let
$$\mathbf{p} = \left[\begin{array}{c} 1900 \\ 1K \end{array} \right], \quad \mathbf{q} = \left[\begin{array}{c} 2000 \\ 100K \end{array} \right].$$

Suppose $\mathbf{p}$ is associated with time 1900 (the year) and $\mathbf{q}$ with 2000. What is the data point for 2050?

2. Consider the parametric line
$$\mathbf{x}(t) = (1 - t) \left[\begin{array}{c} 2 \\ -1 \end{array} \right] + t \left[\begin{array}{c} 0 \\ 1 \end{array} \right].$$

What is its explicit form?

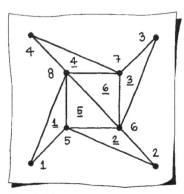

Sketch 21.
A triangulation.

3. Derive the exact definition of a, b, and c in (2.4) from (2.3).

4. Let

$$\mathbf{p} = \begin{bmatrix} 1 \\ 0 \\ 1 \end{bmatrix}, \quad \mathbf{q} = \begin{bmatrix} 0 \\ 1 \\ 1 \end{bmatrix}, \quad \mathbf{r} = \begin{bmatrix} 1 \\ 1 \\ 2 \end{bmatrix}.$$

What is the implicit form of the plane spanned by these points?

5. Suppose you are given a polygon with N points $\mathbf{p}_1, \ldots, \mathbf{p}_N$. Associated with each $\mathbf{p}_i$ is a parameter $u_i = i$. Write pseudocode outlining how you would determine a point on the polygon corresponding to an arbitrary parameter value u.

6. What is the data structure (point, triangle, and neighbor lists) of the triangulation shown in Sketch 21? Follow Example 2.6.

7. If the neighbor list of a triangulation contains no -1 entries, what does that tell us about the triangulation?

8. The process of edge collapsing from Section 2.6 can produce invalid triangulations in some special cases. Describe one such case.

Cubic Bézier Curves   3

Figure 3.1.
A family of cubic Bézier curves and their control polygons.

In this chapter, we cover the basics of cubic Bézier curves. These are the "bread-and-butter" curves of most commercial systems, ranging from CAD/CAM to Graphic Design. While higher order curves are also used, the basic principles are easily explored using this simple curve type. Figure 3.1 shows several Bézier curves which were generated using PostScript language.

3.1 Parametric Curves

Most people know about curves from calculus. There, a curve is a function such as

$$y = 2x - 2x^2,$$

plotted in Sketch 22.

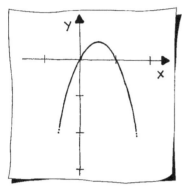

Sketch 22.
The graph of a function.

That sketch does not show the function $y = 2x - 2x^2$; it shows its *graph*. The graph is a set of points

$$\begin{bmatrix} x \\ y \end{bmatrix} = \begin{bmatrix} x \\ 2x - 2x^2 \end{bmatrix}. \tag{3.1}$$

From calculus, we know that there is a horizontal tangent where $\frac{dy}{dx} = 0$, in our example at $x = \frac{1}{2}$.

A *parametric curve* is a more complicated concept, and not all facts from calculus carry over in a straightforward way.

We already encountered a parametric curve in Section 2.1; there, we were dealing with straight lines in parametric form:

$$\begin{bmatrix} x \\ y \end{bmatrix} = \begin{bmatrix} (1-t)a_x + tb_x \\ (1-t)a_y + tb_y \end{bmatrix}.$$

This is the straight line through the points **a** and **b**.

In general, a parametric curve is of the form

$$\begin{bmatrix} x \\ y \end{bmatrix} = \begin{bmatrix} f(t) \\ g(t) \end{bmatrix}. \tag{3.2}$$

In the straight line example, both f and g are linear functions. For a general parametric curve, they can be *any* kind of function. Just as we discussed for parametric lines in Section 2.1, the *domain* of a parametric curve is the real line.

EXAMPLE 3.1

Consider the curve defined by

$$\begin{bmatrix} x \\ y \end{bmatrix} = \begin{bmatrix} t \\ 2t - 2t^2 \end{bmatrix}.$$

This curve is identical to the one given by (3.1).

EXAMPLE 3.2

Let's rotate the curve in Example 3.1 by 90 degrees.[1] Now it is

$$\begin{bmatrix} x \\ y \end{bmatrix} = \begin{bmatrix} -2t + 2t^2 \\ t \end{bmatrix}.$$

---

[1]See Section 1.4 for the rotation matrix.

Checking Sketch 23, we see that the curve's geometry did not change, only its position was affected.

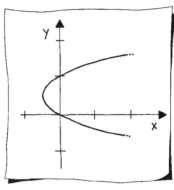

Sketch 23.
A rotated curve.

This example brings out an important difference between parametric curves and the graphs of functions. The concept of "zero slope" or "horizontal tangents" is important for functions; it characterizes extreme points. But for parametric curves, zero slopes do not signify geometric properties; after all, a simple rotation changes horizontal tangents. The geometry of a curve does not change under rotations or other affine maps, yet "horizontal tangents" certainly do!

Finally, parametric curves are defined in 3D in a straightforward manner:

$$\left[\begin{array}{c} x \\ y \\ z \end{array} \right] = \left[\begin{array}{c} f(t) \\ g(t) \\ h(t) \end{array} \right].$$

A simple example is the *helix*. It is given by

$$\mathbf{x} = \left[\begin{array}{c} x \\ y \\ z \end{array} \right] = \left[\begin{array}{c} \cos(t) \\ \sin(t) \\ t \end{array} \right].$$

An illustration is given in Sketch 24.

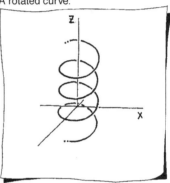

Sketch 24.
A helix.

3.2   Cubic Bézier Curves

In this book, we will not be interested in general parametric curves as the helix above; rather, we will focus on polynomial and piecewise polynomial curves.[2] The most important type of polynomial curves is known as *Bézier curves*, named after Pierre Bézier. While they are defined for any polynomial degree, we will first focus on the cubic case $n = 3$.

EXAMPLE 3.3

We define a parametric curve by setting

$$\mathbf{x}(t) = \left[\begin{array}{c} x \\ y \end{array} \right] = \left[\begin{array}{c} -(1-t)^3 + t^3 \\ 3(1-t)^2 t - 3(1-t)t^2 \end{array} \right].$$

___
[2]Piecewise schemes are introduced in Chapter 9.

From this definition, it is not at all obvious what the curve might look like. Let's rewrite it:

$$\mathbf{x} = \begin{bmatrix} x \\ y \end{bmatrix} = (1-t)^3 \begin{bmatrix} -1 \\ 0 \end{bmatrix} + 3(1-t)^2 t \begin{bmatrix} 0 \\ 1 \end{bmatrix} + 3(1-t)t^2 \begin{bmatrix} 0 \\ -1 \end{bmatrix} + t^3 \begin{bmatrix} 1 \\ 0 \end{bmatrix}.$$

This way of rewriting the polynomial expresses the polynomial in terms of a combination of points. The four points form a polygon, and it roughly resembles the curve segment from $\mathbf{x}(0)$ to $\mathbf{x}(1)$. See Sketch 25. We may compute the point on the curve corresponding to $t = 0.5$ and obtain

$$\mathbf{x}(0.5) = \begin{bmatrix} 0 \\ 0 \end{bmatrix}.$$

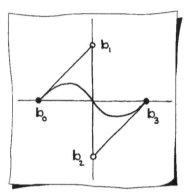

Sketch 25.
A cubic Bézier curve.

In general, we define a *cubic Bézier curve* by

$$\mathbf{x}(t) = (1-t)^3 \mathbf{b}_0 + 3(1-t)^2 t \mathbf{b}_1 + 3(1-t)t^2 \mathbf{b}_2 + t^3 \mathbf{b}_3, \qquad (3.3)$$

where the 2D or 3D points $\mathbf{b}_i$, the *Bézier control points*, form the *Bézier polygon* of the curve. Commonly, the special polynomial expressions in (3.3) are written as

$$\mathbf{x}(t) = B_0^3 \mathbf{b}_0 + B_1^3 \mathbf{b}_1 + B_2^3 \mathbf{b}_2 + B_3^3 \mathbf{b}_3, \qquad (3.4)$$

where the B_i^3 are called the cubic *Bernstein polynomials*. The $\mathbf{b}_i$ are called the *coefficients* of the polynomial $\mathbf{x}$. More on the Bernstein polynomials in Section 4.1.

Cubic Bézier curves have many important properties; we now list some of them; each should be checked with the examples of Figure 3.2.

1. *Endpoint interpolation*: The curve passes through the polygon endpoints: $\mathbf{x}(0) = \mathbf{b}_0$ and $\mathbf{x}(1) = \mathbf{b}_3$.

2. *Symmetry*: The two polygons, $\mathbf{b}_0, \mathbf{b}_1, \mathbf{b}_2, \mathbf{b}_3$ and $\mathbf{b}_3, \mathbf{b}_2, \mathbf{b}_1, \mathbf{b}_0$, describe the same curve; the only thing that changes is the direction of traversal of the parameter.

3. *Invariance under rotations*: If we rotate the control polygon, then the curve is rotated the same way; see the "N" and "Z" Bézier curves in Figure 3.2.

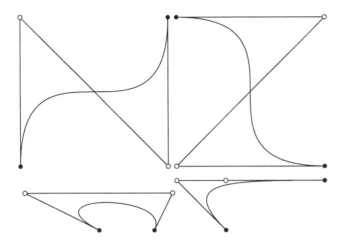

Figure 3.2.
Several cubic Bézier curves.

4. *Invariance under affine maps*: In general, if an affine map is applied to the control polygon, then the curve is mapped by the same map.[3]

5. *Convex hull property*: For $t \in [0, 1]$, the point $\mathbf{x}(t)$ is in the convex hull (see Section 2.4) of the control polygon. See Figure 3.3.

6. *Linear precision*: If the control points $\mathbf{b}_1$ and $\mathbf{b}_2$ are evenly spaced on the straight line between $\mathbf{b}_0$ and $\mathbf{b}_3$, then the cubic Bézier curve is the linear interpolant between $\mathbf{b}_0$ and $\mathbf{b}_3$.

For values of t outside $[0, 1]$, the curve will typically not stay within the control polygon's convex hull. See Figure 3.4 for an illustration. This is called extrapolation. In CAD systems, it is not uncommon to see an extend function which allows a curve to be extended in order to close a gap in geometry. Extending a Bézier curve with extrapolation is not a safe operation due to the unpredictable behavior. Piecewise curve methods are more suitable for this situation; see Chapter 9.

---

[3]Bézier curves are *not* invariant under *projective maps* in this sense!

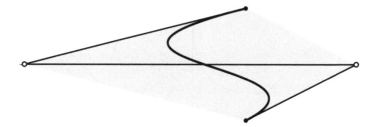

Figure 3.3.
The convex hull property.

Figure 3.4.
A Bézier curve for $t \in [-1, 2]$.

3.3   Derivatives

Differentiating a parametric curve is nothing special: We just differentiate each component. The result—a vector—is the *tangent vector* of
the curve. If we take a Bézier curve of the form (3.3) and differentiate
with respect to the parameter t, we obtain

$$\frac{\mathrm{d}\mathbf{x}(t)}{\mathrm{d}t} = -3(1-t)^2\mathbf{b}_0 + [3(1-t)^2 - 6(1-t)t]\mathbf{b}_1$$
$$+ [6(1-t)t - 3t^2]\mathbf{b}_2 + 3t^2\mathbf{b}_3,$$

and after grouping like terms, this turns into

$$\frac{d\mathbf{x}(t)}{dt} = 3[\mathbf{b}_1 - \mathbf{b}_0](1-t)^2 + 6[\mathbf{b}_2 - \mathbf{b}_1](1-t)t + 3[\mathbf{b}_3 - \mathbf{b}_2]t^2.$$

This is often abbreviated as

$$\frac{d\mathbf{x}(t)}{dt} = 3\Delta\mathbf{b}_0(1-t)^2 + 6\Delta\mathbf{b}_1(1-t)t + 3\Delta\mathbf{b}_2 t^2, \qquad (3.5)$$

where $\Delta\mathbf{b}_i$ is known as the *forward difference*. We will shorten the notation for $d\mathbf{x}(t)/dt$ to $\dot{\mathbf{x}}(t)$.

EXAMPLE 3.4

The derivative of the Bézier curve from Example 3.3 is given by

$$\dot{\mathbf{x}}(t) = 3 \begin{bmatrix} 1 \\ 1 \end{bmatrix} (1-t)^2 + 6 \begin{bmatrix} 0 \\ -2 \end{bmatrix} (1-t)t + 3 \begin{bmatrix} 1 \\ 1 \end{bmatrix} t^2.$$

If we evaluate at $t = 0.5$, we obtain

$$\dot{\mathbf{x}}(0.5) = \begin{bmatrix} 1.5 \\ -1.5 \end{bmatrix}.$$

This tangent vector is shown in Sketch 26.

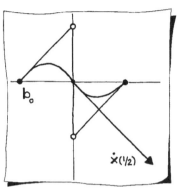

Sketch 26.
A tangent vector.

Another abbreviation: (3.5) can be written as

$$\dot{\mathbf{x}}(t) = 3(\Delta\mathbf{b}_0 B_0^2 + \Delta\mathbf{b}_1 B_1^2 + \Delta\mathbf{b}_2 B_2^2), \qquad (3.6)$$

where the B_i^2 are the quadratic (degree $n = 2$) Bernstein basis functions.[4] Recall that the derivative of a cubic curve is a quadratic curve. For parametric curves, evaluating a "derivative curve" (3.6) produces vectors rather than points.

Take a moment to appreciate the simple form of the derivative of a cubic. The coefficients are the difference vectors of the polygon, scaled by the degree 3. These coefficients are multiplied by the quadratic basis functions—one degree lower than the original curve.

For any Bézier curve, the two most important tangent vectors are those at the curve's endpoints:

$$\dot{\mathbf{x}}(0) = 3\Delta\mathbf{b}_0, \qquad \dot{\mathbf{x}}(1) = 3\Delta\mathbf{b}_2. \qquad (3.7)$$

---

[4]The middle term is 3 rather than 6 because $B_1^2 = 2(1-t)t$.

Thus the control polygon is tangent to the curve at the endpoints—a property that comes in handy when piecing together several Bézier curves!

Higher order derivatives are discussed in Section 4.2.

3.4   The de Casteljau Algorithm

The de Casteljau algorithm is probably the most important algorithm of all of CAGD. Paul de Faget de Casteljau invented it in 1959.

The de Casteljau algorithm is a recursive algorithm that constructs the point $\mathbf{x}(t)$ on a Bézier curve. It proceeds as follows.

Given: $\mathbf{b}_0, \ldots, \mathbf{b}_3$ and a parameter value t.

Task: Find $\mathbf{x}(t)$.

Compute:

$$\mathbf{b}_0^1 = (1 - t)\mathbf{b}_0 + t\mathbf{b}_1$$
$$\mathbf{b}_1^1 = (1 - t)\mathbf{b}_1 + t\mathbf{b}_2$$
$$\mathbf{b}_2^1 = (1 - t)\mathbf{b}_2 + t\mathbf{b}_3.$$

Next:

$$\mathbf{b}_0^2 = (1 - t)\mathbf{b}_0^1 + t\mathbf{b}_1^1$$
$$\mathbf{b}_1^2 = (1 - t)\mathbf{b}_1^1 + t\mathbf{b}_2^1,$$

and finally

$$\mathbf{b}_0^3 = (1 - t)\mathbf{b}_0^2 + t\mathbf{b}_1^2.$$

Then $\mathbf{b}_0^3$ is the desired point on the curve. Sketch 27 illustrates. Notice that this algorithm consists simply of repeated linear interpolation!

In order to see why this is true, simply insert the corresponding terms: In the definition of $\mathbf{b}_0^3$, replace the two terms $\mathbf{b}_0^2$ and $\mathbf{b}_1^2$ by their definitions. Then replace the $\mathbf{b}_i^1$ terms by their definitions, and the result is (3.3)!

A convenient schematic tool for describing the algorithm is to arrange the involved points in a triangular diagram:

$$
\begin{array}{llll}
\mathbf{b}_0 & & & \\
\mathbf{b}_1 & \mathbf{b}_0^1 & & \\
\mathbf{b}_2 & \mathbf{b}_1^1 & \mathbf{b}_0^2 & \\
\mathbf{b}_3 & \mathbf{b}_2^1 & \mathbf{b}_1^2 & \mathbf{b}_0^3.
\end{array}
\tag{3.8}
$$

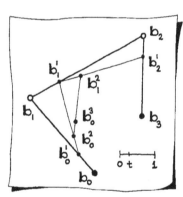

Sketch 27.
The de Casteljau algorithm.

EXAMPLE 3.5

We will evaluate the Bézier curve from (3.3) at $t = 0.5$. The corresponding scheme is:

$$\begin{bmatrix} -1.0 \\ 0.0 \end{bmatrix}$$

$$\begin{bmatrix} 0.0 \\ 1.0 \end{bmatrix} \quad \begin{bmatrix} -0.5 \\ 0.5 \end{bmatrix}$$

$$\begin{bmatrix} 0.0 \\ -1.0 \end{bmatrix} \quad \begin{bmatrix} 0.0 \\ 0.0 \end{bmatrix} \quad \begin{bmatrix} -0.25 \\ 0.25 \end{bmatrix}$$

$$\begin{bmatrix} 1.0 \\ 0.0 \end{bmatrix} \quad \begin{bmatrix} 0.5 \\ -0.5 \end{bmatrix} \quad \begin{bmatrix} 0.25 \\ -0.25 \end{bmatrix} \quad \begin{bmatrix} 0.0 \\ 0.0 \end{bmatrix}$$

thus showing that

$$\mathbf{x}(0.5) = \begin{bmatrix} 0.0 \\ 0.0 \end{bmatrix}.$$

In the implementation of the de Casteljau algorithm, it is not necessary to use a 2D array to simulate the triangular diagram; a 1D array of control points is sufficient. For example, $\mathbf{b}_0^1$ is calculated and loaded into $\mathbf{b}_0$. Notice that $\mathbf{b}_0$ is never needed again in this evaluation, so overwriting this memory is okay. Of course, the original control polygon must be saved somewhere so another evaluation can take place!

We may use the de Casteljau algorithm to evaluate many points on a Bézier curve; all of these operations, when traced graphically, are shown in Figure 3.5.

The de Casteljau algorithm has many practical and theoretical ramifications, as we shall see soon. One involves the computation of derivatives. Inspection of Sketch 27 suggests that the straight line $\overline{\mathbf{b}_0^2 \mathbf{b}_1^2}$ is tangent to the curve. This is indeed so:

$$\dot{\mathbf{x}}(t) = 3[\mathbf{b}_1^2 - \mathbf{b}_0^2]. \tag{3.9}$$

Thus, the derivative is essentially a byproduct of point evaluation! This makes the de Casteljau algorithm very attractive computationally for all situations where $\dot{\mathbf{x}}(t)$ is needed together with $\mathbf{x}(t)$.[5]

---

[5] Proving that (3.9) and (3.6) are equivalent is beyond the scope of this text.

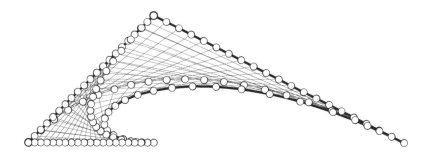

Figure 3.5.
Several examples of the de Casteljau algorithm.

3.5   Subdivision

When you look at Sketch 27, you observe not only the control polygon $\mathbf{b}_0, \mathbf{b}_1, \mathbf{b}_2, \mathbf{b}_3$, but also two more polygons:

$$\mathbf{b}_0, \mathbf{b}_0^1, \mathbf{b}_0^2, \mathbf{b}_0^3 \quad \text{and} \quad \mathbf{b}_0^3, \mathbf{b}_1^2, \mathbf{b}_2^1, \mathbf{b}_3. \qquad (3.10)$$

It turns out that each of these polygons defines the two segments of the curve corresponding to $[0, t]$ and $[t, 1]$ with respect to the original curve.[6] Finding these two polygons is called *subdivision*. In the schematic triangular diagram (3.8), the control points for these curves are along the diagonal and the base of the triangle.

The most important special case is subdivision at $t = 0.5$. Then the curve is split into two arcs at the parametric midpoint. Note that, in general, the two arcs are not of equal length!

Subdivision may be repeated: Each of the two new control polygons may be subdivided, and so on. The resulting sequence of control polygons will ultimately converge to the curve. Convergence is fast, and thus repeated subdivision could be used to plot a curve. Figure 3.6 illustrates. In the highly curved areas, three subdivision steps were necessary to capture the shape of the curve. In the flatter area, only two subdivisions were necessary.

Another application of subdivision is the intersection of a 2D Bézier curve with a line. A *minmax box* is the smallest rectangle with sides parallel to the coordinate axes that contain the curve's control polygon. The curve will also be contained in this box because of the convex

---

[6]The two segments are often referred to as the "left" and "right" segments.

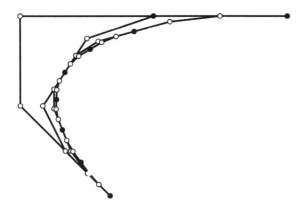

Figure 3.6.
Repeated subdivision for evaluation.

hull property of Bézier curves. Figure 3.7 illustrates the procedure.
As the algorithm proceeds, smaller and smaller minmax boxes are
constructed, focusing on the final intersection points. Note that the
algorithm finds *all* intersections.

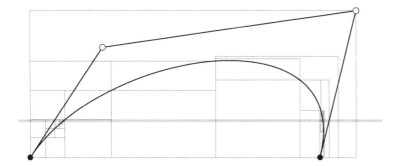

Figure 3.7.
Finding an intersection of a cubic Bézier curve with the x−axis (gray).

The intersection algorithm proceeds iteratively as follows:

0. Find the minmax box of the control polygon.

1. If there is no intersection between the minmax box and the line, exit.

2a. If there is an intersection between the minmax box and the line *and* the minmax box is smaller than a given tolerance, record the center of the minmax box as an intersection point.

2b. Else, subdivide the curve at $t = 0.5$ into two segments and go to Step 0 for each of the resulting control polygons.

3.6 Exploring the Properties of Bézier Curves

In this section, we will study some special Bézier curves—they will help highlight some of the important properties of this type of curve.

EXAMPLE 3.6

Let a Bézier curve be given by

$$\mathbf{b}_0 = \begin{bmatrix} 0 \\ 0 \end{bmatrix}, \quad \mathbf{b}_1 = \begin{bmatrix} 1.5 \\ 1 \end{bmatrix}, \quad \mathbf{b}_2 = \begin{bmatrix} -0.5 \\ 1 \end{bmatrix}, \quad \mathbf{b}_3 = \begin{bmatrix} 1 \\ 0 \end{bmatrix}.$$

It is shown in Figure 3.8. This Bézier curve has a *loop*; i.e., it self-intersects.

EXAMPLE 3.7

Let a Bézier curve be given by

$$\mathbf{b}_0 = \begin{bmatrix} 0 \\ 0 \end{bmatrix}, \quad \mathbf{b}_1 = \begin{bmatrix} 0.7 \\ 1 \end{bmatrix}, \quad \mathbf{b}_2 = \begin{bmatrix} 0.3 \\ 1 \end{bmatrix}, \quad \mathbf{b}_3 = \begin{bmatrix} 1 \\ 0 \end{bmatrix}.$$

It is shown in Figure 3.9. In that figure, two extra points on the curve are marked: These are *inflection points*[7]. A cubic with *two* inflection points? That does not happen for functions, but for parametric cubics, it is possible!

---

[7]For a precise definition see Section 8.2

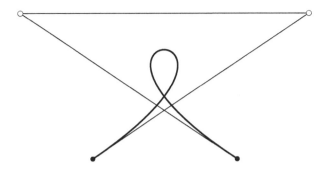

Figure 3.8.
A Bézier curve with a loop.

Figure 3.9.
A Bézier curve with two inflection points.

Figure 3.10.
A Bézier curve with a cusp.

EXAMPLE 3.8

Next, let a Bézier curve be given by

$$\mathbf{b}_0 = \begin{bmatrix} 0 \\ 0 \end{bmatrix}, \quad \mathbf{b}_1 = \begin{bmatrix} 1 \\ 1 \end{bmatrix}, \quad \mathbf{b}_2 = \begin{bmatrix} 0 \\ 1 \end{bmatrix}, \quad \mathbf{b}_3 = \begin{bmatrix} 1 \\ 0 \end{bmatrix}.$$

It is shown in Figure 3.10. At $t = 0.5$, the curve has a *cusp*. These are points where the first derivative vector vanishes.

The next example continues the investigation of our cusp curve.

EXAMPLE 3.9

Referring to the previous example, let us subdivide that curve at $t = 0.5$. The corresponding de Casteljau algorithm is given by:

$$\begin{bmatrix} 0 \\ 0 \end{bmatrix}$$

$$\begin{bmatrix} 1 \\ 1 \end{bmatrix} \quad \begin{bmatrix} 0.5 \\ 0.5 \end{bmatrix}$$

$$\begin{bmatrix} 0 \\ 1 \end{bmatrix} \quad \begin{bmatrix} 0.5 \\ 1 \end{bmatrix} \quad \begin{bmatrix} 0.5 \\ 0.75 \end{bmatrix}$$

$$\begin{bmatrix} 1 \\ 0 \end{bmatrix} \quad \begin{bmatrix} 0.5 \\ 0.5 \end{bmatrix} \quad \begin{bmatrix} 0.5 \\ 0.75 \end{bmatrix} \quad \begin{bmatrix} 0.5 \\ 0.75 \end{bmatrix}$$

The Bézier points $\hat{\mathbf{b}}_i$ of the segment corresponding to $t \in [0, 0.5]$ of the original curve are thus given by

$$\hat{\mathbf{b}}_0 = \begin{bmatrix} 0 \\ 0 \end{bmatrix}, \quad \hat{\mathbf{b}}_1 = \begin{bmatrix} 0.5 \\ 0.5 \end{bmatrix}, \quad \hat{\mathbf{b}}_2 = \begin{bmatrix} 0.5 \\ 0.75 \end{bmatrix}, \quad \hat{\mathbf{b}}_3 = \begin{bmatrix} 0.5 \\ 0.75 \end{bmatrix}.$$

The last two of these Bézier points are identical, hence the cusp.

3.7   The Matrix Form and Monomials

As a preparation for what is to follow, let us rewrite (3.3) using the formalism of dot products. It then becomes

$$\mathbf{b}(t) = \begin{bmatrix} \mathbf{b}_0 & \mathbf{b}_1 & \mathbf{b}_2 & \mathbf{b}_3 \end{bmatrix} \begin{bmatrix} (1-t)^3 \\ 3(1-t)^2 t \\ 3(1-t)t^2 \\ t^3 \end{bmatrix},$$

or taking advantage of the shorthand basis function notation in (3.4)

$$\mathbf{b}(t) = \begin{bmatrix} \mathbf{b}_0 & \mathbf{b}_1 & \mathbf{b}_2 & \mathbf{b}_3 \end{bmatrix} \begin{bmatrix} B_0^3(t) \\ B_1^3(t) \\ B_2^3(t) \\ B_3^3(t) \end{bmatrix}. \qquad (3.11)$$

This is the matrix form of a Bézier curve.

Polynomials were traditionally thought of as combinations of the *monomial polynomials* or *monomials*; they are $1, t, t^2, t^3$ for the cubic case. Equation (3.3) may be rewritten in this form:

$$\mathbf{b}(t) = \mathbf{b}_0 + 3t(\mathbf{b}_1 - \mathbf{b}_0) + 3t^2(\mathbf{b}_2 - 2\mathbf{b}_1 + \mathbf{b}_0) + t^3(\mathbf{b}_3 - 3\mathbf{b}_2 + 3\mathbf{b}_1 - \mathbf{b}_0). \quad (3.12)$$

This allows a more concise formulation using matrices:

$$\mathbf{b}(t) = \begin{bmatrix} \mathbf{b}_0 & \mathbf{b}_1 & \mathbf{b}_2 & \mathbf{b}_3 \end{bmatrix} \begin{bmatrix} 1 & -3 & 3 & -1 \\ 0 & 3 & -6 & 3 \\ 0 & 0 & 3 & -3 \\ 0 & 0 & 0 & 1 \end{bmatrix} \begin{bmatrix} 1 \\ t \\ t^2 \\ t^3 \end{bmatrix}. \quad (3.13)$$

Equation (3.13) shows how to write a Bézier curve in monomial form. A curve in monomial form looks like this:

$$\mathbf{b}(t) = \mathbf{a}_0 + \mathbf{a}_1 t + \mathbf{a}_2 t^2 + \mathbf{a}_3 t^3.$$

Rewritten using the dot product form, this becomes

$$\mathbf{b}(t) = \begin{bmatrix} \mathbf{a}_0 & \mathbf{a}_1 & \mathbf{a}_2 & \mathbf{a}_3 \end{bmatrix} \begin{bmatrix} 1 \\ t \\ t^2 \\ t^3 \end{bmatrix}.$$

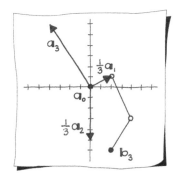

Sketch 28.
The geometric meaning of the monomial form.

Thus the monomial coefficients $\mathbf{a}_i$ are defined as

$$\begin{bmatrix} \mathbf{a}_0 & \mathbf{a}_1 & \mathbf{a}_2 & \mathbf{a}_3 \end{bmatrix} = \begin{bmatrix} \mathbf{b}_0 & \mathbf{b}_1 & \mathbf{b}_2 & \mathbf{b}_3 \end{bmatrix} \begin{bmatrix} 1 & -3 & 3 & -1 \\ 0 & 3 & -6 & 3 \\ 0 & 0 & 3 & -3 \\ 0 & 0 & 0 & 1 \end{bmatrix}. \quad (3.14)$$

Reviewing (3.12), it becomes apparent that the monomial form $\mathbf{a}_i$ have a different geometric interpretation than the Bézier form's $\mathbf{b}_i$. Sketch 28 illustrates that $\mathbf{a}_0$ is a point, however $\mathbf{a}_1, \mathbf{a}_2$, and $\mathbf{a}_3$ are vectors defining the derivatives of the cubic curve at $\mathbf{a}_0$.

The inverse process is not hard either: Given a curve in monomial form, how can we write it as a Bézier curve? Simply rearrange (3.14) to solve for the $\mathbf{b}_i$:

$$\begin{bmatrix} \mathbf{b}_0 & \mathbf{b}_1 & \mathbf{b}_2 & \mathbf{b}_3 \end{bmatrix} = \begin{bmatrix} \mathbf{a}_0 & \mathbf{a}_1 & \mathbf{a}_2 & \mathbf{a}_3 \end{bmatrix} \begin{bmatrix} 1 & -3 & 3 & -1 \\ 0 & 3 & -6 & 3 \\ 0 & 0 & 3 & -3 \\ 0 & 0 & 0 & 1 \end{bmatrix}^{-1}.$$

A matrix inversion is all that is needed here!

Notice that the square matrix in this equation is nonsingular. Because of its nonsingularity, we can conclude that any cubic curve can be written in either the Bézier or the monomial form.

3.8   Exercises

1. Describe what happens if a Bézier curve is *constant*, i.e., all Bézier points are the same.

2. Sketch the five points associated with

$$t = 0, \quad 1/4, \quad 1/2, \quad 3/4, \quad 1$$

   on the three cubic Bézier curves

$$\mathbf{b}_0 = \begin{bmatrix} 0 \\ 0 \end{bmatrix}, \quad \mathbf{b}_1 = \begin{bmatrix} 1 \\ 0 \end{bmatrix}, \quad \mathbf{b}_2 = \begin{bmatrix} 2 \\ 0 \end{bmatrix}, \quad \mathbf{b}_3 = \begin{bmatrix} 3 \\ 0 \end{bmatrix};$$

$$\mathbf{c}_0 = \begin{bmatrix} 0 \\ 0 \end{bmatrix}, \quad \mathbf{c}_1 = \begin{bmatrix} 0 \\ 0 \end{bmatrix}, \quad \mathbf{c}_2 = \begin{bmatrix} 3 \\ 0 \end{bmatrix}, \quad \mathbf{c}_3 = \begin{bmatrix} 3 \\ 0 \end{bmatrix};$$

$$\mathbf{d}_0 = \begin{bmatrix} 0 \\ 0 \end{bmatrix}, \quad \mathbf{d}_1 = \begin{bmatrix} 1.5 \\ 0 \end{bmatrix}, \quad \mathbf{d}_2 = \begin{bmatrix} 1.5 \\ 0 \end{bmatrix}, \quad \mathbf{d}_3 = \begin{bmatrix} 3 \\ 0 \end{bmatrix}.$$

3. Sketch the Bézier polygon

$$\mathbf{b}_0 = \begin{bmatrix} -1 \\ 0 \end{bmatrix}, \quad \mathbf{b}_1 = \begin{bmatrix} -1 \\ 1 \end{bmatrix}, \quad \mathbf{b}_2 = \begin{bmatrix} 1 \\ 1 \end{bmatrix}, \quad \mathbf{b}_3 = \begin{bmatrix} 1 \\ 0 \end{bmatrix}.$$

   Using the triangular diagram, evaluate the Bézier curve at $t = \frac{1}{4}$. What is the derivative of the curve at this point? Add the evaluation point and derivative vector to the sketch.

4. Resketch the polygon from the previous problem. Add to this sketch, using $\diamond$ as a label, the control polygon for the cubic Bézier curve for the segment $[0, \frac{1}{4}]$. Do the same for the curve for the segment $[\frac{1}{4}, 1]$ using the label $\heartsuit$.

5. Construct the Bézier control points for the four parametric curves

$$\begin{bmatrix} t \\ B_0^3 \end{bmatrix}, \quad \begin{bmatrix} t \\ B_1^3 \end{bmatrix}, \quad \begin{bmatrix} t \\ B_2^3 \end{bmatrix}, \quad \begin{bmatrix} t \\ B_3^3 \end{bmatrix}$$

   defined over $[0, 1]$, and sketch each polygon and curve.

6. What is the monomial form of the Bézier curve from Example 3.3?

7. What is the Bézier form of the 3D curve segment corresponding to $t \in [0, 1]$, when the curve is given in monomial form by

$$\mathbf{x}(t) = \begin{bmatrix} t \\ t^2 \\ t^3 \end{bmatrix}?$$

8. Let a Bézier curve be given by

$$\mathbf{b}_0 = \begin{bmatrix} 0 \\ 0 \end{bmatrix}, \quad \mathbf{b}_1 = \begin{bmatrix} 1 \\ 1 \end{bmatrix}, \quad \mathbf{b}_2 = \begin{bmatrix} 1 \\ 1 \end{bmatrix}, \quad \mathbf{b}_3 = \begin{bmatrix} 2 \\ 0 \end{bmatrix}.$$

Sketch the curve for $t \in [-1, 2]$.

Bézier Curves:
Cubic and Beyond

4

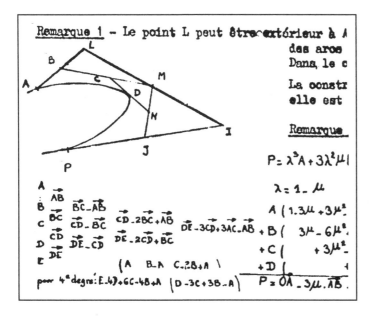

Figure 4.1.

An excerpt from P. de Casteljau's writings.

Bézier curves are not restricted to cubics. Here, we will explore these
more general curves.

4.1 Bézier Curves

A Bézier curve of degree n is defined by

$$\mathbf{x}(t) = \mathbf{b}_0 B_0^n(t) + \mathbf{b}_1 B_1^n(t) + \ldots + \mathbf{b}_n B_n^n(t), \qquad (4.1)$$

where the $B_i^n(t)$ are *Bernstein polynomials*

$$B_i^n(t) = \binom{n}{i}(1-t)^{n-i}t^i. \qquad (4.2)$$

The *binomial coefficients* are defined as

$$\binom{n}{i} = \left\{ \begin{array}{ll} \frac{n!}{i!(n-i)!} & \text{if} \quad 0 \le i \le n \\ 0 & \text{else.} \end{array} \right. \qquad (4.3)$$

In the cubic case, this is identical to (3.4). The Bernstein polynomials of degree four are shown in Figure 4.2. The control polygons in the figure are explained in a more thorough discussion of Bernstein polynomials in Section 4.9.

Several examples of higher degree Bézier curves are shown in Figure 4.3. These examples show how a user might influence the shape of a Bézier curve by adding more control points or moving control points.

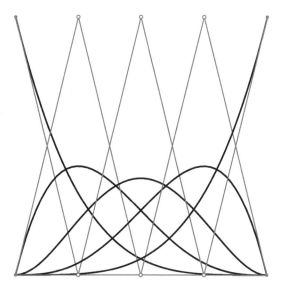

Figure 4.2.

The Bernstein polynomials of degree four plotted over $[0, 1]$.

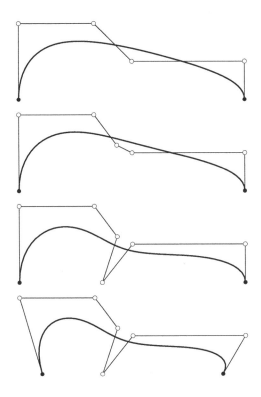

Figure 4.3.

A sequence of Bézier curves.

For general degrees, there are essentially no new properties to report; everything carries over from the cubic case. Be sure to review the properties listed in Section 3.2. Let's revisit the major topics from Chapter 3.

4.2 Derivatives Revisited

For the derivative, we have

$$\dot{\mathbf{x}}(t) = n[\Delta \mathbf{b}_0 B_0^{n-1} + \ldots + \Delta \mathbf{b}_{n-1} B_{n-1}^{n-1}], \qquad (4.4)$$

where $\Delta \mathbf{b}_i = \mathbf{b}_{i+1} - \mathbf{b}_i$. This is again a Bézier curve; its degree is $n-1$ and its coefficients are vectors.

Bézier curves may be differentiated more than once. The kth derivative at parameter value t is given by

$$\frac{d^k \mathbf{x}(t)}{dt^k} = \frac{n!}{(n-k)!}[\Delta^k \mathbf{b}_0 B_0^{n-k}(t) + \ldots + \Delta^k \mathbf{b}_{n-k} B_{n-k}^{n-k}(t)], \quad (4.5)$$

where Δ^k is the forward difference operator which is recursively defined by

$$\Delta^k \mathbf{b}_i = \Delta^{k-1} \mathbf{b}_{i+1} - \Delta^{k-1} \mathbf{b}_i, \quad (4.6)$$

where $\Delta^0 \mathbf{b}_i = \mathbf{b}_i$. Three examples:

$$k = 2: \quad \mathbf{b}_{i+2} - 2\mathbf{b}_{i+1} + \mathbf{b}_i$$
$$k = 3: \quad \mathbf{b}_{i+3} - 3\mathbf{b}_{i+2} + 3\mathbf{b}_{i+1} - \mathbf{b}_i$$
$$k = 4: \quad \mathbf{b}_{i+4} - 4\mathbf{b}_{i+3} + 6\mathbf{b}_{i+2} - 4\mathbf{b}_{i+1} + \mathbf{b}_i$$

Notice that the coefficients oscillate in sign and are identical in value to levels from Pascal's triangle.[1]

At the endpoints, the derivative calculations simplify. With slightly abbreviated notation, we have the kth derivative of $\mathbf{x}(0)$ and $\mathbf{x}(1)$ as

$$\mathbf{x}^{(k)}(0) = \frac{n!}{(n-k)!} \Delta^k \mathbf{b}_0 \quad (4.7)$$

and

$$\mathbf{x}^{(k)}(1) = \frac{n!}{(n-k)!} \Delta^k \mathbf{b}_{n-k}. \quad (4.8)$$

EXAMPLE 4.1

Of the higher order derivatives, the second one, denoted by $\ddot{\mathbf{x}}$, is of particular interest. Let us compute $\ddot{\mathbf{x}}(0)$ for the Bézier curve of Example 3.5. From (4.7) we have

$$\ddot{\mathbf{x}}(0) = 6\Delta^2 \mathbf{b}_0 = 6(\mathbf{b}_2 - 2\mathbf{b}_1 + \mathbf{b}_0).$$

Thus,

$$\ddot{\mathbf{x}}(0) = 6 \left(\begin{bmatrix} 0 \\ -1 \end{bmatrix} - 2 \begin{bmatrix} 0 \\ 1 \end{bmatrix} + \begin{bmatrix} -1 \\ 0 \end{bmatrix} \right) = \begin{bmatrix} -6 \\ -18 \end{bmatrix}.$$

Sketch 29 illustrates.

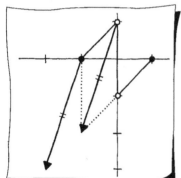

Sketch 29.
A second derivative vector.

---
[1]See Section 4.9 for more on Pascal's triangle.

One nice feature of Bézier curves is the geometric interpretation of the first and second derivatives at the endpoints. The first derivative has the direction of the control polygon leg. The second derivative is in the direction of the diagonal of the parallelogram formed by the three control points at the endpoint. This is illustrated in Sketch 29.

4.3   The de Casteljau Algorithm Revisited

Evaluation of a degree n Bézier curve is done via the de Casteljau algorithm; it is expressed in pseudocode as

for $r = 1, \ldots, n$
   for $i = 0, \ldots, n - r$
$$\mathbf{b}_i^r(t) = (1 - t)\mathbf{b}_i^{r-1} + t\mathbf{b}_{i+1}^{r-1}. \tag{4.9}$$

The point on the curve is given by $\mathbf{x}(t) = \mathbf{b}_0^n(t)$, which is the last point calculated in the double for-loop when $r = n$ and $i = 0$.

The analogue of Figure 3.5 is shown in Figure 4.4, now for a degree four curve. In that figure, you can see several curves formed by circles. Each circle traces out the locus of a $\mathbf{b}_i^r(t)$ as t varies from 0 to 1.

Again, the de Casteljau algorithm *subdivides* the curve into a "left" and a "right" segment. Their respective control polygons are given by

$$\mathbf{b}_0, \mathbf{b}_0^1, \ldots, \mathbf{b}_0^n \quad \text{and} \quad \mathbf{b}_0^n, \mathbf{b}_1^{n-1}, \ldots, \mathbf{b}_n. \tag{4.10}$$

Just as for the cubic curve, the control points for these curves are along the diagonal and the base of the schematic triangular diagram (3.8). Figure 4.5 illustrates a quintic curve subdivided at $t = 3/4$.

The de Casteljau algorithm also provides a way for computing the first derivative:

$$\dot{\mathbf{x}}(t) = n[\mathbf{b}_1^{n-1} - \mathbf{b}_0^{n-1}]. \tag{4.11}$$

Regardless of the degree of the curve, the first derivative vector can be obtained during evaluation by differencing the points in the next to last step in the deCasteljau algorithm.[2] Sketch 30 illustrates the first derivative of a quartic curve at $t = 1/2$.

The second derivative can also be extracted from the de Casteljau algorithm:

$$\ddot{\mathbf{x}}(t) = n(n - 1)[\mathbf{b}_2^{n-2} - 2\mathbf{b}_1^{n-2} + \mathbf{b}_0^{n-2}], \tag{4.12}$$

which is simply a scaling of the second difference of the $(n - 2)^{\text{nd}}$ column in the schematic triangular diagram.

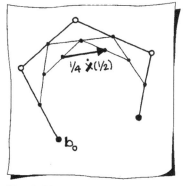

Sketch 30.
A first derivative vector.

---

[2]Proving that (4.11) is equivalent to (4.4) is out of the scope of this text.

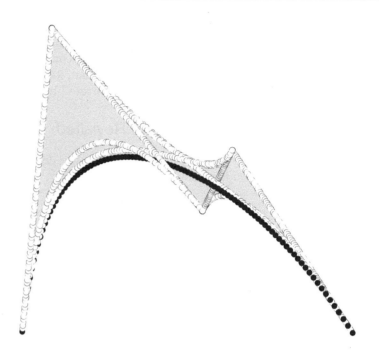

Figure 4.4.
Several de Casteljau algorithm evaluations of a degree four Bézier curve.

4.4   The Matrix Form and Monomials Revisited

Sometimes it is convenient to rewrite (4.1) in *matrix form*. Generalizing the cubics from Section 3.7, we define two vectors N and $\mathbf{B}$ by

$$N = \begin{bmatrix} B_0^n(t) \\ \vdots \\ B_n^n(t) \end{bmatrix}, \qquad \mathbf{B} = \begin{bmatrix} \mathbf{b}_0 \\ \vdots \\ \mathbf{b}_n \end{bmatrix},$$

then (4.1) becomes

$$\mathbf{x}(t) = N^{\mathrm{T}}\mathbf{B}. \tag{4.13}$$

This notation will be useful for dealing with surfaces.

   The matrix notation for a cubic Bézier curve in Section 3.7 proved useful for converting between the Bernstein and monomial basis functions. The matrix in (3.14) gave the relationship between the $\mathbf{b}_i$ and

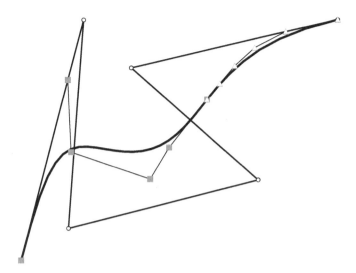

Figure 4.5.
Subdivision of a degree five Bézier curve. The left part's control points are squares and the right part's are triangles.

the $\mathbf{a}_i$. Formulating this matrix for any other degree curve isn't difficult if you notice that the columns of the matrix are scaled forms of the derivative of a Bézier curve at $t = 0$. In other words, form the columns so that

$$\mathbf{a}_0 = \mathbf{b}_0$$

$$\mathbf{a}_i = \frac{\frac{n!}{(n-i)!}\Delta^i \mathbf{b}_0}{i!} = \binom{n}{i}\Delta^i \mathbf{b}_0 \qquad \text{for } i = 1 \ldots n.$$

The Bernstein form is more appealing from a geometric point of view. A curve is defined by control points which mimic the shape of the curve as opposed to the monomial form which is defined in terms of its derivatives. Another advantage of the Bernstein form is that it is numerically more stable than the monomial form.

4.5   Degree Elevation

A degree n polynomial is also one of degree $n+1$, but the leading term has a zero coefficient when the curve is in monomial form. Similarly, we may write a degree n Bézier curve as one of degree $n + 1$. We'll use a quadratic curve,

$$\mathbf{x}(t) = (1 - t)^2 \mathbf{b}_0 + 2(1 - t)t\mathbf{b}_1 + t^2 \mathbf{b}_2,$$

to demonstrate the principle.

The trick is to multiply the quadratic expression by $[t + (1 - t)]$. Notice that $[t + (1 - t)] = 1$! The result of this multiplication:

$$\mathbf{x}(t) = [t(1-t)^2 + (1-t)^3]\mathbf{b}_0 + 2[t^2(1-t) + (1-t)^2 t]\mathbf{b}_1 + [t^3 + t^2(1-t)]\mathbf{b}_2.$$

Reassembling to form barycentric combinations yields

$$\begin{aligned}
\mathbf{x}(t) = &(1 - t)^3 \mathbf{b}_0 \\
&+ 3(1 - t)^2 t[\frac{1}{3}\mathbf{b}_0 + \frac{2}{3}\mathbf{b}_1] \\
&+ 3(1 - t)t^2[\frac{2}{3}\mathbf{b}_1 + \frac{1}{3}\mathbf{b}_2] \\
&+ t^3 \mathbf{b}_2.
\end{aligned}$$

Thus the original quadratic curve may also be written as a cubic with control points

$$\mathbf{x}(t) = B_0^3 \mathbf{b}_0 + B_1^3[\frac{1}{3}\mathbf{b}_0 + \frac{2}{3}\mathbf{b}_1] + B_2^3[\frac{2}{3}\mathbf{b}_1 + \frac{1}{3}\mathbf{b}_2] + B_3^3 \mathbf{b}_2. \quad (4.14)$$

This is then the process of *degree elevation*. The trace or graph of the curve written as a cubic is identical to that of the original quadratic.

EXAMPLE 4.2

Let a quadratic Bézier curve be given by

$$\mathbf{b}_0 = \begin{bmatrix} 0 \\ 0 \end{bmatrix}, \quad \mathbf{b}_1 = \begin{bmatrix} 3 \\ 3 \end{bmatrix}, \quad \mathbf{b}_2 = \begin{bmatrix} 6 \\ 0 \end{bmatrix}$$

Its equation is

$$\mathbf{x}(t) = (1 - t)^2 \mathbf{b}_0 + 2(1 - t)t\mathbf{b}_1 + t^2 \mathbf{b}_2.$$

Degree elevation will allow us to write this curve as

$$\mathbf{x}(t) = (1 - t)^3 \mathbf{c}_0 \qquad\qquad\qquad (4.15)$$
$$+ 3(1 - t)^2 t\mathbf{c}_1 \qquad\qquad\qquad (4.16)$$
$$+ 3(1 - t)t^2 \mathbf{c}_2 \qquad\qquad\qquad (4.17)$$
$$+ t^3 \mathbf{c}_3. \qquad\qquad\qquad (4.18)$$

According to (4.14), the cubic coefficients are

$$\mathbf{c}_0 = \mathbf{b}_0 = \left[\begin{array}{c} 0 \\ 0 \end{array} \right],$$

$$\mathbf{c}_1 = [\frac{1}{3}\mathbf{b}_0 + \frac{2}{3}\mathbf{b}_1] = \left[\begin{array}{c} 2 \\ 2 \end{array} \right],$$

$$\mathbf{c}_2 = [\frac{2}{3}\mathbf{b}_1 + \frac{1}{3}\mathbf{b}_2] = \left[\begin{array}{c} 4 \\ 2 \end{array} \right],$$

$$\mathbf{c}_3 = \mathbf{b}_2 = \left[\begin{array}{c} 6 \\ 0 \end{array} \right].$$

This process of degree elevation is illustrated in Sketch 31.

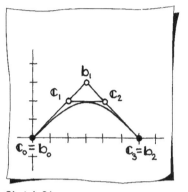

Sketch 31.
Degree elevation from quadratic to cubic.

The above procedure—writing a quadratic Bézier curve as a cubic one—generalizes to higher degrees as well. We then have that a degree n Bézier curve with control polygon $\mathbf{b}_0, \ldots, \mathbf{b}_n$ is identical to one of degree $n+1$ with control polygon $\mathbf{c}_0, \ldots, \mathbf{c}_{n+1}$:

$$\mathbf{c}_0 = \mathbf{b}_0,$$
$$\vdots$$
$$\mathbf{c}_i = \frac{i}{n+1}\mathbf{b}_{i-1} + (1 - \frac{i}{n+1})\mathbf{b}_i, \qquad (4.19)$$
$$\vdots$$
$$\mathbf{c}_{n+1} = \mathbf{b}_n.$$

This process may be written as a matrix operation:

$$\left[\begin{array}{ccccc} 1 & & & & \\ \star & \star & & & \\ & \star & \star & & \\ & & \ddots & \ddots & \\ & & & \star & \star \\ & & & & 1 \end{array} \right] \left[\begin{array}{c} \mathbf{b}_0 \\ \vdots \\ \mathbf{b}_n \end{array} \right] = \left[\begin{array}{c} \mathbf{c}_0 \\ \vdots \\ \mathbf{c}_{n+1} \end{array} \right]. \qquad (4.20)$$

Abbreviated:

$$D\mathbf{B} = \mathbf{C}, \qquad (4.21)$$

D being a matrix with $n + 2$ rows and $n + 1$ columns.

EXAMPLE 4.3

For $n = 2$, the matrix form of the degree elevation process in (4.20) is given by

$$
\begin{bmatrix}
1 & 0 & 0 \\
1/3 & 2/3 & 0 \\
0 & 2/3 & 1/3 \\
0 & 0 & 1
\end{bmatrix}
\begin{bmatrix}
\mathbf{b}_0 \\
\mathbf{b}_1 \\
\mathbf{b}_2
\end{bmatrix}
=
\begin{bmatrix}
\mathbf{c}_0 \\
\mathbf{c}_1 \\
\mathbf{c}_2 \\
\mathbf{c}_3
\end{bmatrix}.
$$

Of course, it is unlikely that you would form such a sparse matrix explicitly; see Section 4.6 for its real purpose.

The process of degree elevation may be applied repeatedly. The resulting sequence of control polygons will converge to the curve, as shown in Figure 4.6. Convergence is too slow for practical purposes, however.

4.6   Degree Reduction

For many applications, the inverse process of *degree reduction* is even more important. Some CAD systems allow degrees up to 40, others only use degree three. Reducing a degree 40 curve to a cubic is not trivial, but the following process will do the job. In practice, several degree three segments will be needed, involving an interplay between subdivision and degree reduction.

In degree reduction, we seek to *approximate* a degree $n + 1$ curve by one of degree n. In terms of (4.21), this means that we would be given $\mathbf{C}$ and wish to find $\mathbf{B}$. Clearly this is not possible in terms of solving a linear system, since D refuses to be a square matrix.

A trick will help: simply multiply both sides of (4.21) by D^{T}, thus getting

$$D^{\mathrm{T}} D \mathbf{B} = D^{\mathrm{T}} \mathbf{C}. \tag{4.22}$$

Now we have a linear system for the unknown $\mathbf{B}$ with a square coefficient matrix $D^{\mathrm{T}} D$—and any linear system solver will do the job!

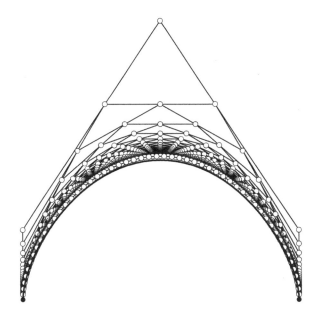

Figure 4.6.
A degree four Bézier curve, degree elevated up to degree 50.

The matrix $D^{\mathrm{T}}D$ does not depend on the given data, just on n. If you have to solve many degree reduction problems, it may therefore be a good idea to store the $L - U$ factorization of $D^{\mathrm{T}}D$.[3]

EXAMPLE 4.4

Let's revisit Example 4.2. Now we will start with the cubic $\mathbf{c}_i$, and use the degree reduction process to find a quadratic approximation. The coefficient matrix of the linear system for the degree reduction process from degree three to degree two is given by

$$D^{\mathrm{T}}D = \frac{1}{9} \left[\begin{array}{ccc} 10 & 2 & 0 \\ 2 & 8 & 2 \\ 0 & 2 & 10 \end{array} \right].$$

---

[3]This is a numerical matrix technique.

The right-hand side of (4.22):

$$D^{\mathrm{T}}C = \frac{1}{3} \begin{bmatrix} 2 & 2 \\ 12 & 8 \\ 22 & 2 \end{bmatrix},$$

where the first column corresponds to the x-components and the second column corresponds to the y-components. Each problem, in x and y, is sent separately to the linear system solver. The solution matrix B is

$$\begin{bmatrix} 0 & 0 \\ 3 & 3 \\ 6 & 0 \end{bmatrix},$$

which makes

$$\mathbf{b}_0 = \begin{bmatrix} 0 \\ 0 \end{bmatrix}, \quad \mathbf{b}_1 = \begin{bmatrix} 3 \\ 3 \end{bmatrix}, \quad \mathbf{b}_2 = \begin{bmatrix} 6 \\ 0 \end{bmatrix}.$$

The solution of (4.22) will, in general, not pass through the original curve endpoints $\mathbf{c}_0$ and $\mathbf{c}_{n+1}$; this could simply be enforced after solving the linear system. We get *endpoint interpolation* in the example above because the cubic was a degree elevated quadratic!

4.7   Bézier Curves over General Intervals

Just as discussed for linear interpolation in Section 2.1, we might want to associate a Bézier curve with the parameter interval $[a, b]$ rather than $[0, 1]$. Again, let u be the global parameter associated with the interval $[a, b]$, then (2.2) describes the *parameter transformation* to find the local parameter t. The local parameter is needed for evaluation via (4.9) to obtain the curve segment that runs from $\mathbf{b}_0$ to $\mathbf{b}_n$. Note that the trace of the Bézier curve is the same regardless of the parameter interval associated with it.

If we plugged the global parameter into the de Casteljau algorithm, we would *extrapolate*, as was discussed in Section 3.2 and illustrated in Figure 3.4.

4.8   Functional Bézier Curves

Recall the discussion of graphs of functions versus parametric curves in Section 3.1. A parametric curve, whose form is given in (3.2), was shown to be more general than a functional curve. In fact, (3.1) illustrates that the graph of a functional curve can be thought of as a parametric curve of the form

$$
\begin{bmatrix} x \\ y \end{bmatrix} = \begin{bmatrix} x(t) \\ y(t) \end{bmatrix} = \begin{bmatrix} t \\ g(t) \end{bmatrix};
$$

one dimension is restricted to be a linear polynomial. Another name for a functional curve is a *nonparametric curve*.

Let's investigate how to write a (polynomial) functional curve in Bézier form. For now, let's restrict t to be within the interval $[0, 1]$, or $t \in [0, 1]$. First of all, the polynomial function $g(t)$ would take the form

$$
g(t) = b_0 B_0^n + \ldots + b_n B_n^n,
$$

where the b_i are scalar values, or *Bézier ordinates*. It now remains to write the linear polynomial t as a degree n polynomial to match the degree of $g(t)$. From Section 3.2, we know that Bézier curves have *linear precision*, and the linear interpolant written as a degree n polynomial requires the control points to be evenly spaced.[4] In this instance, the *abscissa* values are evenly spaced. Thus, the functional Bézier curve takes the form

$$
\mathbf{b}(t) = \begin{bmatrix} 0 \\ b_0 \end{bmatrix} B_0^n \ldots + \begin{bmatrix} j/n \\ b_j \end{bmatrix} B_j^n \ldots + \begin{bmatrix} 1 \\ b_n \end{bmatrix} B_n^n.
$$

A functional Bézier curve is illustrated in Figure 4.7.

If we would like to construct the function so that $t \in [a, b]$, the abscissa values become

$$
a + j\frac{(b - a)}{n} \quad j = 0, \ldots, n.
$$

4.9   More on Bernstein Polynomials

Equations (4.2) and (4.3) in Section 4.1 introduced the general form for *Bernstein polynomials*. Let's look a little closer at these polynomials, in turn understanding the behavior of Bézier curves.

The de Casteljau algorithm is generally preferred for the evaluation of Bézier curves, rather than directly computing the Bernstein polynomials. Thus, to plot the Bernstein polynomials, keep in mind that

---

[4]Compare this with the degree elevation formula (4.19) applied to the Bézier curve $(1 - t)0 + t1$.

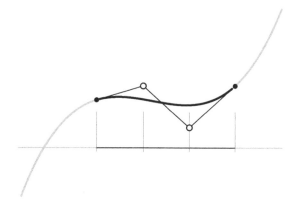

Figure 4.7.
A cubic polynomial written as a functional cubic Bézier curve.

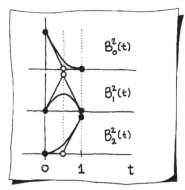

Sketch 32.
Quadratic Bernstein polynomials.

they are functions (see Section 4.8) of t, and then formulate them as Bézier curves! Sketch 32 illustrates the quadratic Bernstein polynomials; the polynomial B_i^2 is given by evenly spaced abcissae over $[0, 1]$, and the ith control ordinate has value one while all others have value zero. The Bézier points for the degree four basis functions are illustrated in Figure 4.2.

The binomial coefficients (4.3) look complicated, however they present us with another occurrence of *Pascal's triangle*.[5] For degree n, the binomial coefficients are taken from the $(n + 1)^{\text{st}}$ row in Pascal's triangle:

$$
\begin{array}{ccccccccccc}
 & & & & & 1 & & & & & \\
 & & & & 1 & & 1 & & & & \\
 & & & 1 & & 2 & & 1 & & & \\
 & & 1 & & 3 & & 3 & & 1 & & \\
 & 1 & & 4 & & 6 & & 4 & & 1 & \\
1 & & 5 & & 10 & & 10 & & 5 & & 1.
\end{array}
$$

Each element in a row is generated by adding the two elements in the previous row which lie above the element.

The Bernstein polynomials are also called Bernstein *basis functions*. The monomials are another example of basis functions. A set of polynomials of degree n that form a basis allow you to write any

---

[5] Recall these coefficients appeared in Section 4.2.

polynomial of degree less than or equal to n in terms of a unique combination of the basis functions.

By examining Sketch 32 or Figure 4.2, you might notice that for any particular value of t, the sum of the Bernstein polynomials is one:

$$B_0^n + \ldots + B_n^n = 1. \qquad (4.23)$$

A fancy name for this property: forming a *partition of unity*. This is a useful identity to keep in mind when debugging a program! See Chapter 14 for more hints.

The Bernstein polynomials have the property that each is nonnegative within the interval $[0, 1]$. This property, along with the fact that they sum to one, is what gives rise to Bézier curves the *convex hull property*, as discussed in Section 3.2.

Have you noticed the symmetry in the Bernstein polynomials? In terms of an identity:

$$B_i^n(t) = B_{n-i}^n(1 - t).$$

This follows directly from (4.2). This is reflected in Bézier curves by the symmetry property (mentioned in Section 3.2). In other words: It does not matter if we number the Bézier points from "left to right" or from "right to left." In each case, we obtain the same curve geometry.

4.10 Exercises

1. Copy Figure 4.4 from the web site or make a copy some other way. (Go to http://www.farinhansford.com/books/essentials-cagd/ essbook-downloads.html and download Postscript Figures, a zip file of Postscript files. Figure 4.4 is called decassol.ps.) Then mark the trajectory of the points $\mathbf{b}_1^1(t)$, assuming that $\mathbf{b}_0$ is the leftmost control point.

2. Repeat the above for the points $\mathbf{b}_1^3(t)$.

3. What is the computation count of the de Casteljau algorithm?

4. Compute $\ddot{\mathbf{x}}(1)$ for the Bézier curve of Example 4.1.

5. Let a cubic Bézier curve be given by

$$\mathbf{b}_0 = \begin{bmatrix} -4 \\ 0 \\ 0 \end{bmatrix}, \quad \mathbf{b}_1 = \begin{bmatrix} 0 \\ 4 \\ 0 \end{bmatrix}, \quad \mathbf{b}_2 = \begin{bmatrix} 0 \\ -4 \\ 0 \end{bmatrix}, \quad \mathbf{b}_3 = \begin{bmatrix} 4 \\ 0 \\ 0 \end{bmatrix}.$$

Degree elevate this curve to degree four, i.e., compute the new control points c_i, for $i = 0, \ldots, 4$.

6. Degree reduce the cubic Bézier curve from the last problem to degree two.

7. Write the graph of the function $y = x^2 + 2x$ as a Bézier curve over the interval $[0, 1]$. Then repeat with the interval $[-1, 1]$.

8. Let a Bézier curve be defined over the global parameter interval $u \in [-2, 2]$. What is the local parameter t corresponding to $u = -1$?

Putting Curves to Work

5

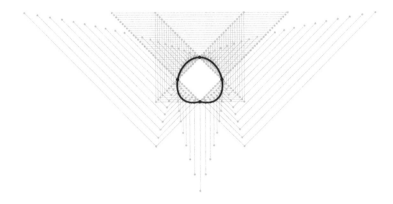

Figure 5.1.
An interpolating polynomial curve, evaluated at fourty points. Intermediate steps in the computations are also shown.

In this chapter, we will see the main uses of parametric curves: describing geometric shapes using methods such as interpolation and approximation.

5.1 Cubic Interpolation

Suppose you are given four points $\mathbf{p}_0, \mathbf{p}_1, \mathbf{p}_2, \mathbf{p}_3$ and you wish to pass a curve through them, just like the situation shown in Sketch 33. There, the points are 2D, but they might as well be 3D. This is called *interpolation*.

We may choose among many kinds of curves; for right now, we'll use a cubic Bézier curve. Every point on a Bézier curve has a parameter

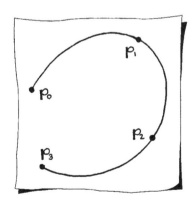

Sketch 33.
A cubic curve through four given points.

value; in order to solve our problem, we have to assign a parameter value t_i to every $\mathbf{p}_i$. A natural choice is to associate each $\mathbf{p}_i$ with a parameter value $t_i = i/3$. Many other suitable choices exist;[1] more on this topic in Section 5.4.

Now our interpolation problem becomes:

Given four point/parameter pairs $(\mathbf{p}_i, t_i)$, find a cubic Bézier curve $\mathbf{x}(t)$ such that

$$\mathbf{x}(t_i) = \mathbf{p}_i; \quad i = 0, 1, 2, 3. \tag{5.1}$$

This simply states that we want the Bézier curve to pass through the data points at the right parameter values.

The desired Bézier curve will be of the form

$$\mathbf{x}(t) = B_0^3(t)\mathbf{b}_0 + B_1^3(t)\mathbf{b}_1 + B_2^3(t)\mathbf{b}_2 + B_3^3(t)\mathbf{b}_3.$$

Writing out all interpolation conditions (5.1) yields

$$\mathbf{p}_0 = B_0^3(t_0)\mathbf{b}_0 + B_1^3(t_0)\mathbf{b}_1 + B_2^3(t_0)\mathbf{b}_2 + B_3^3(t_0)\mathbf{b}_3,$$

$$\mathbf{p}_1 = B_0^3(t_1)\mathbf{b}_0 + B_1^3(t_1)\mathbf{b}_1 + B_2^3(t_1)\mathbf{b}_2 + B_3^3(t_1)\mathbf{b}_3,$$

$$\mathbf{p}_2 = B_0^3(t_2)\mathbf{b}_0 + B_1^3(t_2)\mathbf{b}_1 + B_2^3(t_2)\mathbf{b}_2 + B_3^3(t_2)\mathbf{b}_3,$$

$$\mathbf{p}_3 = B_0^3(t_3)\mathbf{b}_0 + B_1^3(t_3)\mathbf{b}_1 + B_2^3(t_3)\mathbf{b}_2 + B_3^3(t_3)\mathbf{b}_3.$$

These are four equations in the four unknowns $\mathbf{b}_0, \ldots, \mathbf{b}_3$. In order to find a solution, it helps to write them in matrix form:

$$\begin{bmatrix} \mathbf{p}_0 \\ \mathbf{p}_1 \\ \mathbf{p}_2 \\ \mathbf{p}_3 \end{bmatrix} = \begin{bmatrix} B_0^3(t_0) & B_1^3(t_0) & B_2^3(t_0) & B_3^3(t_0) \\ B_0^3(t_1) & B_1^3(t_1) & B_2^3(t_1) & B_3^3(t_1) \\ B_0^3(t_2) & B_1^3(t_2) & B_2^3(t_2) & B_3^3(t_2) \\ B_0^3(t_3) & B_1^3(t_3) & B_2^3(t_3) & B_3^3(t_3) \end{bmatrix} \begin{bmatrix} \mathbf{b}_0 \\ \mathbf{b}_1 \\ \mathbf{b}_2 \\ \mathbf{b}_3 \end{bmatrix}. \tag{5.2}$$

We further abbreviate this as

$$\mathbf{P} = M\mathbf{B}. \tag{5.3}$$

The solution is now simply

$$\mathbf{B} = M^{-1}\mathbf{P}. \tag{5.4}$$

While this looks like the solution to *one* linear system, it is really the solution to two or three systems, depending on the dimensionality of the $\mathbf{p}_i$. An example should clarify:

---

[1] We must require that the t_i are increasing and no two t_i are the same.

EXAMPLE 5.1

Let the $\mathbf{p}_i$ be given by

$$\mathbf{p}_0 = \begin{bmatrix} -1 \\ 0 \end{bmatrix}, \quad \mathbf{p}_1 = \begin{bmatrix} 0 \\ 1 \end{bmatrix}, \quad \mathbf{p}_2 = \begin{bmatrix} 0 \\ -1 \end{bmatrix}, \quad \mathbf{p}_3 = \begin{bmatrix} 1 \\ 0 \end{bmatrix},$$

and set $t_i = i/3$. Then the matrix M for our linear system becomes

$$M = \frac{1}{27} \begin{bmatrix} 27 & 0 & 0 & 0 \\ 8 & 12 & 6 & 1 \\ 1 & 6 & 12 & 8 \\ 0 & 0 & 0 & 27 \end{bmatrix}.$$

We send M and, successively, the two right-hand side vectors,

$$\begin{bmatrix} -1 \\ 0 \\ 0 \\ 1 \end{bmatrix} \quad \text{and} \quad \begin{bmatrix} 0 \\ 1 \\ -1 \\ 0 \end{bmatrix}$$

to a linear system solver, which results in the two solution vectors

$$\begin{bmatrix} -1 \\ 7/6 \\ -7/6 \\ 1 \end{bmatrix} \quad \text{and} \quad \begin{bmatrix} 0 \\ 9/2 \\ -9/2 \\ 0 \end{bmatrix}.$$

Thus, the Bézier points for the interpolating cubic are

$$\mathbf{b}_0 = \begin{bmatrix} -1 \\ 0 \end{bmatrix}, \quad \mathbf{b}_1 = \begin{bmatrix} 7/6 \\ 9/2 \end{bmatrix}, \quad \mathbf{b}_2 = \begin{bmatrix} -7/6 \\ -9/2 \end{bmatrix}, \quad \mathbf{b}_3 = \begin{bmatrix} 1 \\ 0 \end{bmatrix},$$

This example is outlined in Sketch 34.

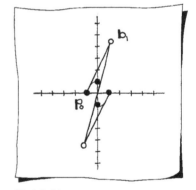

Sketch 34.
Cubic Bézier interpolation.

Note that if you explicitly computed M^{-1}, you may apply it to any number of cubic interpolation problems, just as long as you keep the choice of the t_i unchanged!

5.2 Interpolation Beyond Cubics

Polynomial interpolation also works when more than four data points are given. We would then have data points $\mathbf{p}_0, \ldots, \mathbf{p}_n$ and corresponding parameter values $t_0, \ldots, t_n$. The interpolation problem again leads to a linear system

$$\mathbf{P} = M\mathbf{B}; \tag{5.5}$$

now M is an $(n+1) \times (n+1)$ matrix with elements

$$m_{i,j} = B_j^n(t_i).$$

Again, it is solved using any linear system solver.

While polynomial interpolation is guaranteed to work, it does not produce satisfying results for higher degrees. Figure 5.2 should be convincing enough. The top data set are simply 16 points read off a semicircle, spaced at equal angle increments. The bottom data set has one change: The gray data point now has an x-coordinate of 0.1025, whereas the correct value (supplied in the top part) is 0.1045. Thus a small change in data can lead to large changes in the interpolating curve. Processes with this property are called ill-conditioned.

We used the Bézier form of the interpolating curve here. Different polynomial forms will give the identical result. For example, we might write the interpolating curve in monomial form:

$$\mathbf{x}(t) = \mathbf{a}_0 + \mathbf{a}_1 t + \ldots + \mathbf{a}_n t^n.$$

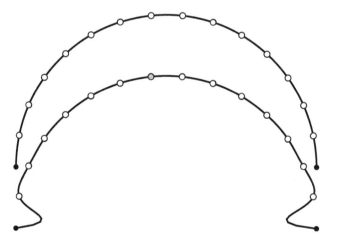

Figure 5.2.

Top: data from a circle; bottom: one point slightly modified.

Now the unknowns are the coefficients $\mathbf{a}_i$. This interpolation problem also leads to a linear system, written as

$$\mathbf{P} = M\mathbf{A};$$

now M is an $(n+1) \times (n+1)$ matrix with elements

$$m_{i,j} = t_i^j.$$

The unknowns $\mathbf{a}_i$ are collected in $\mathbf{A}$. The curve resulting from this system is the same as the one resulting from (5.5).

Another way of writing the interpolating polynomial curve is by using *Lagrange polynomials*. These are defined by

$$L_i^n(t) = \frac{(t - t_0)\dots(t - t_{i-1})(t - t_{i+1})\dots(t - t_n)}{(t_i - t_0)\dots(t_i - t_{i-1})(t_i - t_{i+1})\dots(t_i - t_n)}, \qquad (5.6)$$

and allow a very direct form for the interpolant:

$$\mathbf{x}(t) = L_0^n(t)\mathbf{p}_0 + \dots + L_n^n(t)\mathbf{p}_n. \qquad (5.7)$$

Because the data points appear explicitly, this is called the *cardinal form* of the interpolant to data points and parameters. Notice that in the numerator and denominator of the ith Lagrange polynomial, the $(* - t_i)$ term is missing.

5.3 Aitken's Algorithm

This is a recursive algorithm to compute points on the interpolating polynomial curve; it has some of the characteristics of the de Casteljau algorithm. It is best explained for the cubic case. Assume that we have (somehow) found a quadratic curve $\mathbf{p}_0^2(t)$ through $\mathbf{p}_0, \mathbf{p}_1,$ and $\mathbf{p}_2$ as well as another one, $\mathbf{p}_1^2(t)$ through $\mathbf{p}_1, \mathbf{p}_2,$ and $\mathbf{p}_3$. From these two curves, it is possible to find the interpolating cubic by simple linear interpolation:

$$\mathbf{p}_0^3(t) = \frac{t_3 - t}{t_3 - t_0}\mathbf{p}_0^2(t) + \frac{t - t_0}{t_3 - t_0}\mathbf{p}_1^2(t). \qquad (5.8)$$

Sketch 35 illustrates.

It is easy to verify that (5.8) does indeed interpolate to all four data points. Let's first check $\mathbf{p}_0$:

$$\mathbf{p}_0^3(t_0) = \frac{t_3 - t_0}{t_3 - t_0}\mathbf{p}_0^2(t_0) + \frac{t_0 - t_0}{t_3 - t_0}\mathbf{p}_1^2(t_0) = \mathbf{p}_0.$$

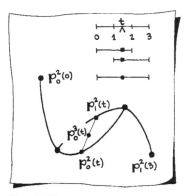

Sketch 35.
Aitken's algorithm.

Interpolation at $\mathbf{p}_3$ is established in the same way. Now for $\mathbf{p}_1$: We simply observe that the factors in (5.8) sum to one. Since both $\mathbf{p}_0^2(t_1) = \mathbf{p}_1$ and $\mathbf{p}_1^2(t_1) = \mathbf{p}_1$, it then follows that also $\mathbf{p}_0^3(t_1) = \mathbf{p}_1$. For $\mathbf{p}_2$, an analogous argument holds.

This begs a question, namely how did we find the quadratic interpolants to begin with? The same process works again:

$$\mathbf{p}_0^2(t) = \frac{t_2 - t}{t_2 - t_0}\mathbf{p}_0^1(t) + \frac{t - t_0}{t_2 - t_0}\mathbf{p}_1^1(t), \tag{5.9}$$

$$\mathbf{p}_1^2(t) = \frac{t_3 - t}{t_3 - t_1}\mathbf{p}_1^1(t) + \frac{t - t_1}{t_3 - t_1}\mathbf{p}_2^1(t). \tag{5.10}$$

In this equation, new terms $\mathbf{p}_i^1$ are introduced, but they are simply linear interpolants of the data. For example,

$$\mathbf{p}_1^1(t) = \frac{t_2 - t}{t_2 - t_1}\mathbf{p}_1 + \frac{t - t_1}{t_2 - t_1}\mathbf{p}_2.$$

Just as in the de Casteljau algorithm, it is convenient to arrange the intermediate points in a triangular array:

$$\begin{array}{llll}
\mathbf{p}_0 & & & \\
\mathbf{p}_1 & \mathbf{p}_0^1 & & \\
\mathbf{p}_2 & \mathbf{p}_1^1 & \mathbf{p}_0^2 & \\
\mathbf{p}_3 & \mathbf{p}_2^1 & \mathbf{p}_1^2 & \mathbf{p}_0^3.
\end{array}$$

The points in the left-most column are given (as well as a parameter value for each point). Aitken's algorithm computes the points in each successive column, and the point on the curve is $\mathbf{p}_0^3$.

See Sketch 36 for the geometry of this computation. The parameter spans in the steps of Aitken's algorithm, such as in (5.8), can seem pretty complicated at first. However, there is an easy way to keep track of this. As depicted in Sketch 36 the first step of the algorithm where the $\mathbf{p}_i^1$ are computed involves spans in the parameter space of length one. Consider all such spans, and with respect to each, determine where the current t-value resides. The span itself corresponds to the input $\mathbf{p}_i$, $\mathbf{p}_{i+1}$ points. In the next step, where the $\mathbf{p}_i^2$ are computed, we now use spans of length two. The points of Step 2 are calculated using the points from Step 1.

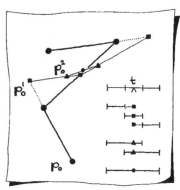

Sketch 36.
The intermediate steps in Aitken's algorithm.

EXAMPLE 5.2

We will evaluate the interpolating cubic through the four data points

$$\mathbf{p}_0 = \begin{bmatrix} -1 \\ 0 \end{bmatrix}, \quad \mathbf{p}_1 = \begin{bmatrix} 0 \\ 1 \end{bmatrix}, \quad \mathbf{p}_2 = \begin{bmatrix} 0 \\ -1 \end{bmatrix}, \quad \mathbf{p}_3 = \begin{bmatrix} 1 \\ 0 \end{bmatrix}$$

with parameter values $(t_0, t_1, t_2, t_3) = (0, 1, 2, 3)$, at $t = 1.5$. The corresponding scheme is:

$$
\begin{bmatrix} -1 \\ 0 \end{bmatrix}
$$
$$
\begin{bmatrix} 0 \\ 1 \end{bmatrix} \quad \begin{bmatrix} 0.5 \\ 1.5 \end{bmatrix}
$$
$$
\begin{bmatrix} 0 \\ -1 \end{bmatrix} \quad \begin{bmatrix} 0 \\ 0 \end{bmatrix} \quad \begin{bmatrix} 0.125 \\ 0.375 \end{bmatrix}
$$
$$
\begin{bmatrix} 1 \\ 0 \end{bmatrix} \quad \begin{bmatrix} -0.5 \\ -1.5 \end{bmatrix} \quad \begin{bmatrix} -0.125 \\ -0.375 \end{bmatrix} \quad \begin{bmatrix} 0 \\ 0 \end{bmatrix},
$$

hence

$$
\mathbf{p}_0^3(1.5) = \begin{bmatrix} 0 \\ 0 \end{bmatrix}.
$$

A sampling of the computation of the intermediate points:

$$
\mathbf{p}_0^1 = -0.5\mathbf{p}_0 + 1.5\mathbf{p}_1
$$
$$
\mathbf{p}_0^2 = 0.25\mathbf{p}_0^1 + 0.75\mathbf{p}_1^1
$$
$$
\mathbf{p}_0^3 = 0.5\mathbf{p}_0^2 + 0.5\mathbf{p}_1^2.
$$

The symmetry in this example makes the other combinations easy to guess.

Aitken's algorithm works for degrees other than cubic. For an nth degree interpolating curve, we will generate intermediate points $\mathbf{p}_i^r$, computed by

$$
\mathbf{p}_i^r(t) = \frac{t_{i+r} - t}{t_{i+r} - t_i}\mathbf{p}_i^{r-1}(t) + \frac{t - t_i}{t_{i+r} - t_i}\mathbf{p}_{i+1}^{r-1}(t) \qquad (5.11)
$$

for $r = 1, \ldots, n$ and $i = 0, \ldots, n - r$. The general form (5.11) of Aitken's algorithm uses *linear interpolation* between two points $\mathbf{p}_i^{r-1}$ and $\mathbf{p}_{i+1}^{r-1}$ over the parameter interval $[t_{i+r}, t_i]$. This may be viewed as an affine map of this interval onto the line through $\mathbf{p}_i^{r-1}$ and $\mathbf{p}_{i+1}^{r-1}$. (This point was illustrated in Sketch 36).

Figure 5.1 illustrates Aitken's algorithm when executed for fourty evaluations. All intermediate points $\mathbf{p}_i^r$ are shown in that figure.

Compare with the corresponding Figure 4.4 for the de Casteljau algorithm!

Did you notice that every data point is involved in the calculation of a point on the interpolating curve? Polynomial interpolation is a *global* operation, as we see in the formulation (5.5).

5.4 Approximation

In many applications, one is given more data points than should be interpolated by a polynomial curve: recall from Section 5.2 that higher degree interpolation becomes ill-conditioned. In such cases, an *approximating* curve will be needed. Such a curve does not pass through the data points exactly; rather it passes near them, still capturing the shape suggested by the given points. The technique best known for finding such curves is known as *least squares approximation*. Figure 5.3 illustrates.

We are now given $l + 1$ data points $\mathbf{p}_0, \ldots, \mathbf{p}_l$, each $\mathbf{p}_i$ being associated with a parameter value t_i. We wish to find a polynomial curve $\mathbf{x}(t)$ of a given degree n such that the distances $\|\mathbf{p}_i - \mathbf{x}(t_i)\|$ are small. Ideally, we would have $\mathbf{p}_i = \mathbf{x}(t_i); i = 0, \ldots, l$. If our polynomial curve $\mathbf{x}(t)$ is of the form

$$\mathbf{x}(t) = \mathbf{b}_0 B_0^n(t) + \ldots + \mathbf{b}_n B_n^n(t)$$

we would like the following to hold:

$$\mathbf{b}_0 B_0^n(t_0) + \ldots + \mathbf{b}_n B_n^n(t_0) = \mathbf{p}_0$$

$$\vdots$$

$$\mathbf{b}_0 B_0^n(t_l) + \ldots + \mathbf{b}_n B_n^n(t_l) = \mathbf{p}_l.$$

This may be condensed into matrix form:

$$\begin{bmatrix} B_0^n(t_0) & \ldots & B_n^n(t_0) \\ & \vdots & \\ & \vdots & \\ B_0^n(t_l) & \ldots & B_n^n(t_l) \end{bmatrix} \begin{bmatrix} \mathbf{b}_0 \\ \vdots \\ \mathbf{b}_n \end{bmatrix} = \begin{bmatrix} \mathbf{p}_0 \\ \vdots \\ \vdots \\ \mathbf{p}_l \end{bmatrix}. \tag{5.12}$$

Or, even shorter:

$$M\mathbf{B} = \mathbf{P}. \tag{5.13}$$

Since we assume the number l of data points is larger than the degree n of the curve, this linear system is clearly *overdetermined*. We attack it by simply multiplying both sides by M^{T}:

$$M^{\mathrm{T}}M\mathbf{B} = M^{\mathrm{T}}\mathbf{P}. \qquad (5.14)$$

This is a linear system with $n + 1$ equations in $n + 1$ unknowns, with a square and symmetric coefficient matrix $M^{\mathrm{T}}M$. Its solution is straightforward since $M^{\mathrm{T}}M$ is always invertible. The linear system (5.14) is called the system of *normal equations*. The curve $\mathbf{B}$ is the one polynomial of degree n which minimizes the sum of the $\|\mathbf{p}_i - \mathbf{x}(t_i)\|$. Note that any modification of the t_i would result in an entirely different solution.

An example is shown in Figure 5.3. The data set, 79 points, are from a cross section of a wing (it's a noisy data set). It is approximated by a least squares quintic using parameter values selected to reflect the spacing of the data.

The choice of the "right" degree for this type of problem is not easy. Typically, it would be a trial-and-error process.

We used the Bézier representation above, however (5.12) could be rewritten using any other set of basis functions, such as the monomial form.

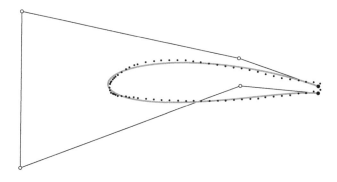

Figure 5.3.
Least squares approximation to a wing. A quintic Bézier curve with uniform parameters assigned to the data. Figure courtesy of Jeong-Jun Song.

5.5   Finding the Right Parameters

In both curve interpolation and approximation, we encountered state-
ments like "given parameter values t_i." In practice, these are not normally given and have to be made up. If there are $l + 1$ points $\mathbf{p}_i$,
an obvious choice is to set $t_i = i/l$; this is known as the *uniform set
of parameters.*

Another, not too involved choice is to arrange the parameters such
that they reflect the spacing of the data points: If the distance
between two points is relatively large, then their parameter values
should also be fairly different. This method is known as *chord length
parameters* and is given by

$$t_0 = 0 \tag{5.15}$$

$$t_1 = t_0 + \|\mathbf{p}_1 - \mathbf{p}_0\| \tag{5.16}$$

$$\vdots \tag{5.17}$$

$$t_l = t_{l-1} + \|\mathbf{p}_l - \mathbf{p}_{l-1}\|. \tag{5.18}$$

When constructing a curve with Bernstein polynomials, it is advan-
tageous to *normalize* the parameters by scaling them to live between
zero and one:

$$t_i = \frac{t_i - t_0}{t_l - t_0}.$$

In general, the chord length parameter selection method is supe-
rior to the uniform method since it takes into account the geometry
of the data. For example, in the case of the interpolation problem
from Figure 5.2, the distorted data set with chord length parameters
generates a curve that is indistinguishable from the true circle.

5.6   Hermite Interpolation

So far, we have discussed curve fitting to points only. Sometimes,
one also knows tangent vectors, and a different kind of interpolation
problem arises, known as *cubic Hermite interpolation.* We are now
given two points $\mathbf{p}_0, \mathbf{p}_1$ and two tangent vectors $\mathbf{v}_0, \mathbf{v}_1$. The objective
is to find a cubic polynomial curve $\mathbf{x}(t)$ that interpolates to these data:

$$\mathbf{x}(0) = \mathbf{p}_0,$$
$$\dot{\mathbf{x}}(0) = \mathbf{v}_0,$$
$$\dot{\mathbf{x}}(1) = \mathbf{v}_1,$$
$$\mathbf{x}(1) = \mathbf{p}_1.$$

This situation is shown in Sketch 37.

We will write $\mathbf{x}$ in cubic Bézier form, and therefore must determine four Bézier points $\mathbf{b}_0, \mathbf{b}_1, \mathbf{b}_2, \mathbf{b}_3$. Two of them are quickly determined:

$$\mathbf{b}_0 = \mathbf{p}_0, \quad \mathbf{b}_3 = \mathbf{p}_1.$$

For the remaining two, we recall (from Section 3.3) the endpoint derivative for Bézier curves:

$$\dot{\mathbf{x}}(0) = 3\Delta\mathbf{b}_0, \quad \dot{\mathbf{x}}(1) = 3\Delta\mathbf{b}_2.$$

We can easily solve for $\mathbf{b}_1$ and $\mathbf{b}_2$:

$$\mathbf{b}_1 = \mathbf{p}_0 + \frac{1}{3}\mathbf{v}_0, \qquad \mathbf{b}_2 = \mathbf{p}_1 - \frac{1}{3}\mathbf{v}_1.$$

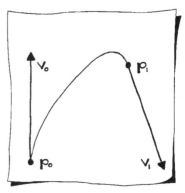

Sketch 37.
Cubic Hermite interpolation.

This solution is shown in Sketch 38.

It is possible to rewrite this result such that the given data appear *explicitly* in the equation for the interpolant. So far, our interpolant is in Bézier form:

$$\mathbf{x}(t) = \mathbf{p}_0 B_0^3(t) + (\mathbf{p}_0 + \frac{1}{3}\mathbf{v}_0)B_1^3(t) + (\mathbf{p}_1 - \frac{1}{3}\mathbf{v}_1)B_2^3(t) + \mathbf{p}_1 B_3^3(t).$$

We simply rearrange:

$$\mathbf{x}(t) = \mathbf{p}_0 H_0^3(t) + \mathbf{v}_0 H_1^3(t) + \mathbf{v}_1 H_2^3(t) + \mathbf{p}_1 H_3^3(t), \qquad (5.19)$$

where we have set

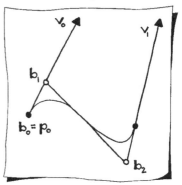

Sketch 38.
Cubic Hermite interpolation and Bézier solution.

$$H_0^3(t) = B_0^3(t) + B_1^3(t),$$
$$H_1^3(t) = \frac{1}{3}B_1^3(t),$$
$$H_2^3(t) = -\frac{1}{3}B_2^3(t),$$
$$H_3^3(t) = B_2^3(t) + B_3^3(t).$$

The H_i^3 are called "cubic Hermite polynomials." Analogous to the Lagrange polynomials, the Hermite polynomials are the *cardinal form* for the interpolant to data points and tangent vectors since the input data appear explicitly.

The length of the tangent vectors $\mathbf{v}_0$ and $\mathbf{v}_1$ are an important factor for the final curve's shape. This length is not very intuitive to a user; we thus recommend against the use of the Hermite form unless exact derivatives are known.

5.7 Exercises

1. Let four points be given by

$$\mathbf{p}_0 = \begin{bmatrix} 0 \\ 0 \end{bmatrix}, \quad \mathbf{p}_1 = \begin{bmatrix} 1 \\ 1 \end{bmatrix}, \quad \mathbf{p}_2 = \begin{bmatrix} 2 \\ 4 \end{bmatrix}, \quad \mathbf{p}_3 = \begin{bmatrix} 3 \\ 9 \end{bmatrix}.$$

   Setting $t_i = i/3$, find the Bézier polygon of the interpolating cubic.

2. Using the same data as in Exercise 1, evaluate the interpolating cubic at $t = 0.5$ using Aitken's algorithm. (Be sure to sketch the spans as well as the intermediate points!) Verify that you get the same answer by applying the de Casteljau algorithm to the Bézier curve from the previous problem.

3. Sketch the interpolant to the three points

$$\begin{bmatrix} -1 \\ 0 \end{bmatrix} \quad \begin{bmatrix} 0 \\ 1 \end{bmatrix} \quad \begin{bmatrix} 1 \\ 0 \end{bmatrix}$$

   with parameters $t_i = (0, 1, 2)$. Use Aitken's algorithm to evaluate at $t = 0.5$ and $t = 1.5$ to guess the shape of the curve. Now repeat this exercise, but with parameters $t_i = (0, 0.25, 2)$. Explain the result.

4. Sketch (manually, if you like) the three quadratic Lagrange polynomials for $(t_0, t_1, t_2) = (0, 4, 5)$.

5. What are the Hermite forms of the three Bézier curves from Section 3.6?

6. The data points of Figure 5.3 are in the file `wing.dat` from the data sets file on the web page `http://www.farinhansford.com/ books/essentials-cagd/essbook-downloads.html`. Find and plot the least squares fit using degree three and seven curves with chord length parameters.

Bézier Patches 6

Figure 6.1.
One of the most famous objects in computer graphics is the "Utah teapot," made up of 32 Bézier patches.

In this chapter, we will encounter surfaces for the first time. We will cover the basic definitions and go on to extend the concept of Bézier curves to surfaces. A famous object which is composed of Bézier patches is the "Utah teapot," shown in Figure 6.1.[1]

6.1 Parametric Surfaces

A parametric curve is the result of a mapping of the real line into 2- or 3-space. A parametric surface is defined in a similar way: It is the result of a map of the real plane into 3-space. This "real plane" is called the *domain* of the surface. It is simply a plane with a

---

[1] Our teapot was created by Mary Zhu using 3D Studio Max.

coordinate system such that every point has coordinates (u, v). The corresponding 3D surface point is then a point:

$$\mathbf{x}(u, v) = \begin{bmatrix} f(u, v) \\ g(u, v) \\ h(u, v) \end{bmatrix}.$$ (6.1)

EXAMPLE 6.1

The parametric surface given by

$$\mathbf{x}(u, v) = \begin{bmatrix} u \\ v \\ u^2 + v^2 \end{bmatrix}$$

is illustrated in Sketch 39. Of course, only a portion of the surface is illustrated; the surface extends infinitely from each edge. This parametric surface happens to be a *functional surface* because two of the coordinate functions in (6.1) are simply u and v.

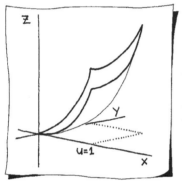

Sketch 39.
A parametric surface.

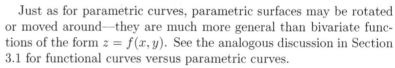

Just as for parametric curves, parametric surfaces may be rotated or moved around—they are much more general than bivariate functions of the form $z = f(x, y)$. See the analogous discussion in Section 3.1 for functional curves versus parametric curves.

6.2   Bilinear Patches

We will typically be interested in a finite piece of a parametric surface; It is the image of a rectangle in the domain.[2] The finite piece of surface will then be called a *patch*.

To get started, we pick a special rectangle: It is the *unit square*, defined by

$$\{(u, v) : 0 \leq u, v \leq 1\}.$$

We map it to a surface patch which is defined by four points $\mathbf{b}_{0,0}, \mathbf{b}_{0,1}, \mathbf{b}_{1,0}, \mathbf{b}_{1,1}$ in a very straightforward way: We simply set

$$\mathbf{x}(u, v) = \begin{bmatrix} 1-u & u \end{bmatrix} \begin{bmatrix} \mathbf{b}_{0,0} & \mathbf{b}_{0,1} \\ \mathbf{b}_{1,0} & \mathbf{b}_{1,1} \end{bmatrix} \begin{bmatrix} 1-v \\ v \end{bmatrix}.$$ (6.2)

---

[2]Shapes more complicated than rectangles are possible, but they are beyond the scope of this book.

This surface patch is linear in both the u and v parameters, hence the name *bilinear patch*.

Equation (6.2) gives a very nice concise expression for the bilinear patch, but it doesn't convey very much geometric information. By simply rewriting the bilinear patch as

$$\mathbf{x}(u, v) = (1 - v)\mathbf{p}^u + v\mathbf{q}^u \tag{6.3}$$

where

$$\mathbf{p}^u = (1 - u)\mathbf{b}_{0,0} + u\mathbf{b}_{1,0} \quad \text{and} \quad \mathbf{q}^u = (1 - u)\mathbf{b}_{0,1} + u\mathbf{b}_{1,1},$$

we can gather a much better feeling for the shape of the bilinear patch.

EXAMPLE 6.2

Let four points $\mathbf{b}_{i,j}$ be given by

$$\mathbf{b}_{0,0} = \begin{bmatrix} 0 \\ 0 \\ 0 \end{bmatrix}, \quad \mathbf{b}_{1,0} = \begin{bmatrix} 1 \\ 0 \\ 0 \end{bmatrix}, \quad \mathbf{b}_{0,1} = \begin{bmatrix} 0 \\ 1 \\ 0 \end{bmatrix}, \quad \mathbf{b}_{1,1} = \begin{bmatrix} 1 \\ 1 \\ 1 \end{bmatrix}.$$

First compute

$$\mathbf{p}^u = 0.75 \begin{bmatrix} 0 \\ 0 \\ 0 \end{bmatrix} + 0.25 \begin{bmatrix} 1 \\ 0 \\ 0 \end{bmatrix} = \begin{bmatrix} 0.25 \\ 0 \\ 0 \end{bmatrix}$$

$$\mathbf{q}^u = 0.75 \begin{bmatrix} 0 \\ 1 \\ 0 \end{bmatrix} + 0.25 \begin{bmatrix} 1 \\ 1 \\ 1 \end{bmatrix} = \begin{bmatrix} 0.25 \\ 1 \\ 0.25 \end{bmatrix}$$

as illustrated in Sketch 40. Then, the point on the patch is given by

$$\mathbf{x}(0.25, 0.5) = 0.5\mathbf{p}^u + 0.5\mathbf{q}^u = \begin{bmatrix} 0.25 \\ 0.5 \\ 0.125 \end{bmatrix}.$$

A rendered image of this patch is shown in Figure 6.2.

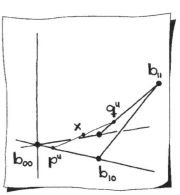

Sketch 40.
A bilinear patch.

Notice that the bilinear patch as expressed in (6.3) could have also been written as

$$\mathbf{x}(u, v) = (1 - u)\mathbf{p}^v + u\mathbf{q}^v \tag{6.4}$$

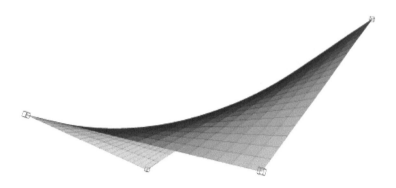

Figure 6.2.
A bilinear patch.

where

$$\mathbf{p}^v = (1 - v)\mathbf{b}_{0,0} + v\mathbf{b}_{0,1} \quad \text{and} \quad \mathbf{q}^v = (1 - v)\mathbf{b}_{1,0} + v\mathbf{b}_{1,1}.$$

Recompute the point on the patch from Example 6.2 to check that you get the same answer! Don't forget to create your own sketch with both computations.

A bilinear patch, also called a *hyperbolic paraboloid*, is covered by two families of straight lines. That's not so easy to see from (6.2), but it is easy to see from (6.3) or (6.4). Consider (parametric) lines in the domain that are parallel to the sides of the unit square. A line constant in u but varying in v would take the form $(\bar{u}, v)$; a line constant in v but varying in u would take the form $(u, \bar{v})$. These two families of lines on the patch correspond to the lines generated by (6.3) and (6.4), respectively. These are called *isoparametric curves* on the patch since only one parameter is allowed to vary. Four special isoparametric curves (lines) on the patch are the edges corresponding to

$$(u, 0), \quad (u, 1), \quad (0, v), \quad (1, v).$$

However, a hyperbolic paraboloid also contains *curves*, as we will now see. Consider the line $u = v$ in the domain, i.e., the diagonal of the unit square. In parametric form (in the domain), it may be written as $u(t) = t, v(t) = t$. This domain diagonal is mapped to the 3D curve

$$\mathbf{d}(t) = \mathbf{x}(t, t)$$

on the surface. In more detail:

$$\mathbf{d}(t) = \begin{bmatrix} 1-t & t \end{bmatrix} \begin{bmatrix} \mathbf{b}_{0,0} & \mathbf{b}_{0,1} \\ \mathbf{b}_{1,0} & \mathbf{b}_{1,1} \end{bmatrix} \begin{bmatrix} 1-t \\ t \end{bmatrix}.$$

Collecting terms now gives

$$\mathbf{d}(t) = (1-t)^2 \mathbf{b}_{0,0} + 2(1-t)t[\frac{1}{2}\mathbf{b}_{0,1} + \frac{1}{2}\mathbf{b}_{1,0}] + t^2 \mathbf{b}_{1,1}. \qquad (6.5)$$

This is a quadratic Bézier curve!

EXAMPLE 6.3

Using the bilinear patch from Example 6.2, let's compute the curve on the surface corresponding to the diagonal in the domain: $u(t) = t, v(t) = t$. From (6.5), we know this is a quadratic curve with Bézier points

$$\mathbf{c}_0 = \mathbf{b}_{0,0} = \begin{bmatrix} 0 \\ 0 \\ 0 \end{bmatrix} \quad \mathbf{c}_1 = \frac{1}{2}[\mathbf{b}_{1,0} + \mathbf{b}_{0,1}] = \begin{bmatrix} 0.5 \\ 0.5 \\ 0 \end{bmatrix} \quad \mathbf{c}_2 = \mathbf{b}_{1,1} = \begin{bmatrix} 1 \\ 1 \\ 1 \end{bmatrix}.$$

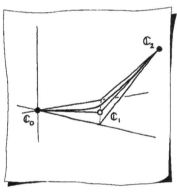

Thus,

$$\mathbf{d}(t) = \mathbf{c}_0 B_0^2(t) + \mathbf{c}_1 B_1^2(t) + \mathbf{c}_2 B_2^2(t)$$

and Sketch 41 illustrates.

Sketch 41.
The diagonal of a bilinear patch.

6.3   Bézier Patches

Let's slightly rewrite the bilinear patch from (6.2) using linear Bernstein polynomials:

$$\mathbf{x}(u, v) = \begin{bmatrix} B_0^1(u) & B_1^1(u) \end{bmatrix} \begin{bmatrix} \mathbf{b}_{0,0} & \mathbf{b}_{0,1} \\ \mathbf{b}_{1,0} & \mathbf{b}_{1,1} \end{bmatrix} \begin{bmatrix} B_0^1(v) \\ B_1^1(v) \end{bmatrix}. \qquad (6.6)$$

This suggests the following generalization:

$$\mathbf{x}(u, v) = \begin{bmatrix} B_0^m(u) & \dots & B_m^m(u) \end{bmatrix} \begin{bmatrix} \mathbf{b}_{0,0} & \dots & \mathbf{b}_{0,n} \\ \vdots & & \vdots \\ \mathbf{b}_{m,0} & \dots & \mathbf{b}_{m,n} \end{bmatrix} \begin{bmatrix} B_0^n(v) \\ \vdots \\ B_n^n(v) \end{bmatrix}.$$

$$(6.7)$$

For $m = n = 1$, we recover the bilinear case; for the special setting $n = m = 3$, we have *bicubic* Bézier patches. Expanding (6.7) would result in

$$\mathbf{x}(u,v) = \mathbf{b}_{0,0}B_0^m(u)B_0^n(v) + \ldots + \mathbf{b}_{i,j}B_i^m(u)B_j^n(v) + \ldots + \mathbf{b}_{m,n}B_m^m(u)B_n^n(v).$$

Equation (6.7) is conveniently abbreviated as

$$\mathbf{x}(u,v) = M^{\mathrm{T}}\mathbf{B}N. \tag{6.8}$$

In this form, it is the surface generalization of the curve equation 4.13).

In evaluating (6.7) at a parameter pair (u,v), it is natural to first multiply the first two factors and then multiply the result with the last factor. We would thus define

$$\mathbf{C} = M^{\mathrm{T}}\mathbf{B} = [\mathbf{c}_0, \ldots, \mathbf{c}_n] \tag{6.9}$$

and then write the final result as

$$\mathbf{x}(u,v) = \mathbf{C}N. \tag{6.10}$$

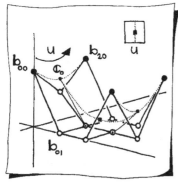

Sketch 42.
Evaluation of a Bézier patch via a $u = $ constant isocurve.

Let's call this the *2-stage explicit* evaluation method. The term 'explicit' refers to the fact that the Bernstein polynomials are explicitly evaluated.

In (6.9), the elements $\mathbf{c}_0, \ldots, \mathbf{c}_n$ of $\mathbf{C}$ do not depend on the parameter value v; thus, after having computed $\mathbf{C}$, we could use it to compute several points $\mathbf{x}(u,v_1), \mathbf{x}(u,v_2), \ldots$. Thus, $\mathbf{C}$ contains the Bézier points which define the curve $\mathbf{C}N$ with constant u and variable v; such a curve is known as an *isoparametric curve*, or isocurve for short. Unlike the bilinear patch in Section 6.2, the isoparametric curves are, in general, not straight lines.

We now give an example, which is shown in Sketch 42.

EXAMPLE 6.4

Let $m = 2$ and $n = 3$; the Bézier patch is given by the control net

$$\mathbf{B} = \begin{bmatrix} \begin{bmatrix} 0 \\ 0 \\ 6 \end{bmatrix} & \begin{bmatrix} 3 \\ 0 \\ 0 \end{bmatrix} & \begin{bmatrix} 6 \\ 0 \\ 0 \end{bmatrix} & \begin{bmatrix} 9 \\ 0 \\ 6 \end{bmatrix} \\ \begin{bmatrix} 0 \\ 3 \\ 3 \end{bmatrix} & \begin{bmatrix} 3 \\ 3 \\ 0 \end{bmatrix} & \begin{bmatrix} 6 \\ 3 \\ 0 \end{bmatrix} & \begin{bmatrix} 9 \\ 3 \\ 0 \end{bmatrix} \\ \begin{bmatrix} 0 \\ 6 \\ 6 \end{bmatrix} & \begin{bmatrix} 3 \\ 6 \\ 0 \end{bmatrix} & \begin{bmatrix} 6 \\ 6 \\ 0 \end{bmatrix} & \begin{bmatrix} 9 \\ 6 \\ 6 \end{bmatrix} \end{bmatrix}$$

We select $(u, v) = (0.5, 0.5)$. First, we compute the quadratic Bernstein basis functions with respect to $u = 0.5$:

$$M^{\mathrm{T}} = \begin{bmatrix} 0.25 & 0.5 & 0.25 \end{bmatrix}.$$

Keep in mind from (4.23) that they sum to one! This results in intermediate control points

$$\mathbf{C} = M^{\mathrm{T}}\mathbf{B} = \begin{bmatrix} \begin{bmatrix} 0 \\ 3 \\ 4.5 \end{bmatrix} & \begin{bmatrix} 3 \\ 3 \\ 0 \end{bmatrix} & \begin{bmatrix} 6 \\ 3 \\ 0 \end{bmatrix} & \begin{bmatrix} 9 \\ 3 \\ 3 \end{bmatrix} \end{bmatrix}.$$

As you can see, the points in $\mathbf{C}$ are the Bézier points of an isoparametric curve containing $\mathbf{x}(0.5, 0.5)$.

Next, we compute the cubic Bernstein basis functions with respect to $v = 0.5$:

$$N = \begin{bmatrix} 0.125 \\ 0.375 \\ 0.375 \\ 0.125 \end{bmatrix}$$

and we have

$$\mathbf{x}(0.5, 0.5) = \mathbf{C}N = \begin{bmatrix} 4.5 \\ 3 \\ 0.9375 \end{bmatrix}$$

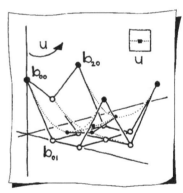

In developing the matrix form of a Bézier patch, we chose to first compute $\mathbf{C} = M^{\mathrm{T}}\mathbf{B}$. Of course, we could have started the other way: Set $\mathbf{D} = \mathbf{B}N$ and then $\mathbf{x} = M^{\mathrm{T}}\mathbf{D}$. The result is the same; Sketch 43 illustrates this alternative approach to Example 6.4 with the v-isoparametric curve labeled $\mathbf{d}_0, \ldots, \mathbf{d}_m$.

Sketch 43.
Evaluation of a Bézier patch via a $v = \text{constant}$ isocurve.

6.4 Properties of Bézier Patches

Bézier patches have many properties that are essentially carbon copies of the curve ones.

1. *Endpoint interpolation*: Analogous to the curve case, the patch passes through the four corner control points, that is

$$\begin{aligned} \mathbf{x}(0,0) &= \mathbf{b}_{0,0} & \mathbf{x}(1,0) &= \mathbf{b}_{m,0} \\ \mathbf{x}(0,1) &= \mathbf{b}_{0,n} & \mathbf{x}(1,1) &= \mathbf{b}_{m,n}. \end{aligned}$$

However, this property is more powerful for the surface case than for the curve case. We also have that control polygon boundaries are the control points of the patch boundary curves. For example: The curve $\mathbf{x}(u,1)$ has the control polygon $\mathbf{b}_{0,n}, \ldots, \mathbf{b}_{m,n}$.

2. *Symmetry*: We could re-index the control net so that any of the corners corresponds to $\mathbf{b}_{0,0}$, and evaluation would result in a patch with the same shape as the original one.

3. *Affine invariance*: Apply an affine map to the control net, and then evaluate the patch. This surface will be identical to the surface created by applying the same affine map to the original patch.

4. *Convex hull property*: For $(u,v) \in [0,1] \times [0,1]$, the patch $\mathbf{x}(u,v)$ is in the convex hull of the the the control net.

5. *Bilinear precision*: Sketch 44 illustrates a degree $m \times n$ patch with boundary control points which are evenly spaced on lines connecting the corner control points, and the interior control points are evenly-spaced on lines connecting boundary control points on adjacent edges. This patch is identical to the bilinear interpolant to the four corner control points.

6. *Tensor product*: Bézier patches are in the class of tensor product surfaces. This property allows Bézier patches to be dealt with in terms of isoparametric curves, which in turn simplifies evaluation and other operations. The breakdown of (6.8) into (6.9) and (6.10) illustrates this point nicely.

The tensor product property is a very powerful conceptual tool for understanding Bézier patches. Sketch 45 illustrates how the shape of a Bézier patch can be thought of as a record of the shape of a template moving through space. This template can change shape as it moves, and its shape and position is guided by "columns" of Bézier control points.

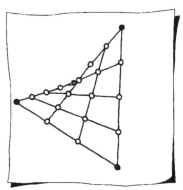

Sketch 44.
A degree 3×4 control net with bilinear precision.

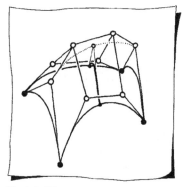

Sketch 45.
Bézier patch as locus of a moving and deforming template.

6.5 Derivatives

Derivatives for parametric curves are tangent vectors. For surfaces, the same is true: A derivative is the tangent vector of a curve on the surface.

To be more specific, let $\mathbf{x}(u,v)$ be a point on a parametric surface (think Bézier patch for simplicity). There are two isoparametric curves through this point; let's focus on the $v = constant$ one, shown in Sketch 46. This is a parametric curve which happens to be restricted to lie on the surface; its parameter is u. We may thus differentiate it with respect to u. The resulting tangent vector $\mathbf{x}_u$ is called a *partial derivative* and is denoted

$$\mathbf{x}_u(u,v) = \frac{\partial \mathbf{x}(u,v)}{\partial u}.$$

It is also called a u-partial. A simple example will explain.

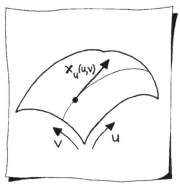

Sketch 46.
A partial.

EXAMPLE 6.5

Let's take the patch from Example 6.4. What is the partial $\mathbf{x}_v(0.5, 0.5)$? The control polygon C for the $u = 0.5$ isoparametric curve is illustrated in Sketch 42. Its derivative curve is

$$\mathbf{x}_v(0.5, v) = 3(\Delta \mathbf{c}_0 B_0^2(v) + \Delta \mathbf{c}_1 B_1^2(v) + \Delta \mathbf{c}_2 B_2^2(v)),$$

where

$$\Delta \mathbf{c}_0 = \begin{bmatrix} 3 \\ 0 \\ -4.5 \end{bmatrix} \quad \Delta \mathbf{c}_1 = \begin{bmatrix} 3 \\ 0 \\ 0 \end{bmatrix} \quad \Delta \mathbf{c}_2 = \begin{bmatrix} 3 \\ 0 \\ 3 \end{bmatrix}.$$

Evaluate this quadratic Bézier curve at $v = 0.5$, and you have

$$\mathbf{x}_v(0.5, 0.5) = \begin{bmatrix} 9 \\ 0 \\ -1.125 \end{bmatrix}.$$

This partial is shown in Sketch 47.

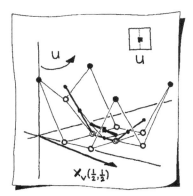

Sketch 47.
A v-partial.

    In an analogous way, we may define u-partials. We just differentiate the isoparametric curve with control points $\mathbf{D} = \mathbf{B}N$.

    Another possibility for computing derivatives is to find a closed-form expression. For it, we differentiate (6.7) with respect to u:

$$\mathbf{x}_u(u,v) =$$

$$m\left[B_0^{m-1}(u) \ \ldots \ B_{m-1}^{m-1}(u)\right] \begin{bmatrix} \Delta^{1,0}\mathbf{b}_{0,0} & \ldots & \Delta^{1,0}\mathbf{b}_{0,n} \\ \vdots & & \vdots \\ \Delta^{1,0}\mathbf{b}_{m-1,0} & \ldots & \Delta^{1,0}\mathbf{b}_{m-1,n} \end{bmatrix} \begin{bmatrix} B_0^n(v) \\ \vdots \\ B_n^n(v) \end{bmatrix}. (6.11)$$

The terms $\Delta^{1,0}\mathbf{b}_{i,j}$ denote forward differences:

$$\Delta^{1,0}\mathbf{b}_{i,j} = \mathbf{b}_{i+1,j} - \mathbf{b}_{i,j}.$$

Thus the closed-form u-partial derivative expression is a degree $(m-1)$ $\times$ n patch with a control net consisting of vectors rather than points.

EXAMPLE 6.6

For the patch from Example 6.4, the u-partial is given by forming differences of the control points of the original patch in the u-direction; see Sketch 48. The u-partial at any (u, v) may be determined by evaluating

$$\mathbf{x}_u(u,v) = 2 \begin{bmatrix} B_0^1(u) & B_1^1(u) \end{bmatrix} \begin{bmatrix} \begin{bmatrix} 0 \\ 3 \\ -3 \end{bmatrix} & \begin{bmatrix} 0 \\ 3 \\ 0 \end{bmatrix} & \begin{bmatrix} 0 \\ 3 \\ 0 \end{bmatrix} & \begin{bmatrix} 0 \\ 3 \\ -6 \end{bmatrix} \\ \begin{bmatrix} 0 \\ 3 \\ 3 \end{bmatrix} & \begin{bmatrix} 0 \\ 3 \\ 0 \end{bmatrix} & \begin{bmatrix} 0 \\ 3 \\ 0 \end{bmatrix} & \begin{bmatrix} 0 \\ 3 \\ 6 \end{bmatrix} \end{bmatrix} \begin{bmatrix} B_0^3(v) \\ B_1^3(v) \\ B_2^3(v) \\ B_3^3(v) \end{bmatrix}$$

Evaluation of this patch at $(u, v) = (0.5, 0.5)$ yields

$$\mathbf{x}_u(0.5, 0.5) = \begin{bmatrix} 0 \\ 6 \\ 0 \end{bmatrix}.$$

Notice that all x- and y-coordinates of the control vectors are identical. This is due to the fact that our initial control net is evenly spaced in both the x- and y-directions. See Section 6.12 for more information on this type of surface.

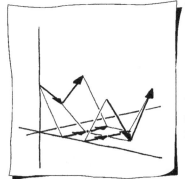

Sketch 48.
Taking differences of a control net.

Let's compare the v-partial derivative computed using the closed-form to the method which isolates an isoparametric curve. For the closed-form v-partial, we differentiate (6.7) with respect to v:

$$\mathbf{x}_v(u,v) =$$

$$n \begin{bmatrix} B_0^m(u) & \dots & B_m^m(u) \end{bmatrix} \begin{bmatrix} \Delta^{0,1}\mathbf{b}_{0,0} & \dots & \Delta^{0,1}\mathbf{b}_{0,n-1} \\ \vdots & & \vdots \\ \Delta^{0,1}\mathbf{b}_{m,0} & \dots & \Delta^{0,1}\mathbf{b}_{m,n-1} \end{bmatrix} \begin{bmatrix} B_0^{n-1}(v) \\ \vdots \\ B_{n-1}^{n-1}(v) \end{bmatrix}. \quad (6.12)$$

The terms $\Delta^{0,1}\mathbf{b}_{i,j}$ denote forward differences:

$$\Delta^{0,1}\mathbf{b}_{i,j} = \mathbf{b}_{i,j+1} - \mathbf{b}_{i,j}.$$

Thus the closed-form v-partial derivative expression is a degree $m \times (n - 1)$ patch.

EXAMPLE 6.7

For the patch from Example 6.4, the v-partial is given by forming differences of the control points of the original patch in the v-direction: $\Delta^{0,1}\mathbf{b}_{i,j}$. The v-partial at any (u,v) may be determined by evaluating

$$
\mathbf{x}_v(u,v) = 3 \begin{bmatrix} B_0^2(u) & B_1^2(u) & B_2^2(u) \end{bmatrix}
\begin{bmatrix}
\begin{bmatrix} 3 \\ 0 \\ -6 \end{bmatrix} & \begin{bmatrix} 3 \\ 0 \\ 0 \end{bmatrix} & \begin{bmatrix} 3 \\ 0 \\ 6 \end{bmatrix} \\[2ex]
\begin{bmatrix} 3 \\ 0 \\ -3 \end{bmatrix} & \begin{bmatrix} 3 \\ 0 \\ 0 \end{bmatrix} & \begin{bmatrix} 3 \\ 0 \\ 0 \end{bmatrix} \\[2ex]
\begin{bmatrix} 3 \\ 0 \\ -6 \end{bmatrix} & \begin{bmatrix} 3 \\ 0 \\ 0 \end{bmatrix} & \begin{bmatrix} 3 \\ 0 \\ 6 \end{bmatrix}
\end{bmatrix}
\begin{bmatrix} B_0^2(v) \\ B_1^2(v) \\ B_2^2(v) \end{bmatrix}.
$$

Evaluation of this patch at $(u,v) = (0.5, 0.5)$ yields the same result as in Example 6.5.

6.6   Higher Order Derivatives

A Bézier patch may be differentiated several times. The resulting derivatives of order k are the kth partials of the patch. Focusing on v-partials for now, we obtain

$$\mathbf{x}_v^{(k)}(u,v) =$$

$$
\frac{n!}{(n-k)!} \begin{bmatrix} B_0^m(u) & \dots & B_m^m(u) \end{bmatrix}
\begin{bmatrix}
\Delta^{0,k}\mathbf{b}_{0,0} & \dots & \Delta^{0,k}\mathbf{b}_{0,n-k} \\
\vdots & & \vdots \\
\Delta^{0,k}\mathbf{b}_{m,0} & \dots & \Delta^{0,k}\mathbf{b}_{m,n-k}
\end{bmatrix}
\begin{bmatrix}
B_0^{n-k}(v) \\
\vdots \\
B_{n-k}^{n-k}(v)
\end{bmatrix}
\tag{6.13}
$$

The terms $\Delta^{0,k}\mathbf{b}_{i,j}$ are k^{th} forward differences in the v-direction, acting only on the second subscripts; recall the definition from (4.6). Of course, the kth u-partial follows similarly.

Another commonly used derivative is the mixed partial, or *twist vector*. It is denoted by $\mathbf{x}_{u,v}(u,v)$ and is obtained in either of two ways:

$$\mathbf{x}_{u,v}(u,v) = \frac{\partial \mathbf{x}_u(u,v)}{\partial v} \quad \text{or} \quad \frac{\partial \mathbf{x}_v(u,v)}{\partial u}.$$

If we take the second option, we simply differentiate (6.12) with respect to u, which amounts to taking differences in the u-direction:

$$\mathbf{x}_{u,v}(u,v) =$$

$$mn \left[B_0^{m-1}(u) \dots B_{m-1}^{m-1}(u) \right] \begin{bmatrix} \Delta^{1,1}\mathbf{b}_{0,0} & \dots & \Delta^{1,1}\mathbf{b}_{0,n-1} \\ \vdots & & \vdots \\ \Delta^{1,1}\mathbf{b}_{m-1,0} & \dots & \Delta^{1,1}\mathbf{b}_{m-1,n-1} \end{bmatrix} \begin{bmatrix} B_0^{n-1}(v) \\ \vdots \\ B_{n-1}^{n-1}(v) \end{bmatrix} \quad (6.14)$$

The terms $\Delta^{1,1}\mathbf{b}_{i,j}$ are obtained by differencing the control net first in the u-direction

$$\Delta^{1,1}\mathbf{b}_{i,j} = \Delta^{0,1}(\mathbf{b}_{i+1,j} - \mathbf{b}_{i,j}) = \Delta^{0,1}\mathbf{b}_{i+1,j} - \Delta^{0,1}\mathbf{b}_{i,j},$$

and then in the v-direction. They are explicitly given by

$$\Delta^{1,1}\mathbf{b}_{i,j} = \mathbf{b}_{i+1,j+1} - \mathbf{b}_{i+1,j} - \mathbf{b}_{i,j+1} + \mathbf{b}_{i,j}. \quad (6.15)$$

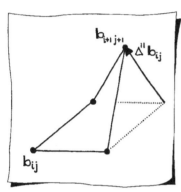

Sketch 49.
A twist coefficient.

These coefficients have a simple geometric meaning: They indicate how much the quadrilateral defined by the four points in (6.15) deviates from a parallelogram. See Sketch 49.

EXAMPLE 6.8

Revisiting Example 6.2, we compute

$$\mathbf{x}_{u,v}(u,v) = \Delta^{1,1}\mathbf{b}_{0,0} = \begin{bmatrix} 0 \\ 0 \\ 1 \end{bmatrix},$$

since $B_0^0(u) = 1$ for all u. Thus, a bilinear patch has a *constant* twist vector.

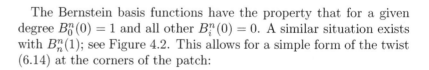

The Bernstein basis functions have the property that for a given degree $B_0^n(0) = 1$ and all other $B_i^n(0) = 0$. A similar situation exists with $B_n^n(1)$; see Figure 4.2. This allows for a simple form of the twist (6.14) at the corners of the patch:

$$\mathbf{x}_{u,v}(0,0) = mn\Delta^{1,1}\mathbf{b}_{0,0} \qquad \mathbf{x}_{u,v}(0,1) = mn\Delta^{1,1}\mathbf{b}_{0,n-1} \quad (6.16)$$

$$\mathbf{x}_{u,v}(1,0) = mn\Delta^{1,1}\mathbf{b}_{m-1,0} \qquad \mathbf{x}_{u,v}(1,1) = mn\Delta^{1,1}\mathbf{b}_{m-1,n-1}. \quad (6.17)$$

6.7   The de Casteljau Algorithm

When we evaluated a patch in Section 6.3, we defined an intermediate set of points, constituting $\mathbf{C} = M^{\mathrm{T}}\mathbf{B}$. Writing out the computation for each individual $\mathbf{c}_i$ yields

$$\mathbf{c}_0 = B_0^m(u)\mathbf{b}_{0,0} + \ldots + B_m^m(u)\mathbf{b}_{m,0},$$
$$\mathbf{c}_1 = B_0^m(u)\mathbf{b}_{0,1} + \ldots + B_m^m(u)\mathbf{b}_{m,1},$$
$$\ldots$$
$$\mathbf{c}_n = B_0^m(u)\mathbf{b}_{0,n} + \ldots + B_m^m(u)\mathbf{b}_{m,n}.$$

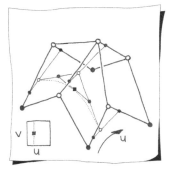

Each of these computations is the evaluation of a degree m Bézier curve. Instead of using the explicit Bernstein polynomial form, we might also use the de Casteljau algorithm for each of the $\mathbf{c}_i$.[3] The resulting geometry is shown in Sketch 50.

The final evaluation step, $\mathbf{x}(u,v) = \mathbf{C}N$, is again the evaluation of a Bézier curve. This time it is of degree n. Again, we may use the de Casteljau algorithm. Thus, the de Casteljau algorithm for Bézier patches, let's call this the *2-stage de Casteljau* evaluation method, simply consists of repeated calls to the de Casteljau algorithm for curves.

Sketch 50.
Evaluating a patch using de Casteljau algorithms.

The advantage of this geometric approach is that it allows computation of a derivative along with computation of a point. Once we have the control polygon $\mathbf{C}$, then we evaluate point and tangent just as described for the curve case in Section 3.4. This tangent, for the algorithm above would correspond to $\mathbf{x}_v$. It even has another advantage, as outlined in Section 6.8.

The roles of u and v can be switched in the 2-stage de Casteljau evaluation method. We can also evaluate a Bézier patch by first computing $\mathbf{D} = \mathbf{B}N$ and then $\mathbf{x} = M^{\mathrm{T}}\mathbf{D}$. The tangent to the curve with control polygon $\mathbf{D}$ results in $\mathbf{x}_u$.

6.8   Normals

The *normal vector*, or *normal*, is a fundamental geometric concept which is used throughout computer graphics and CAD/CAM. At a given point $\mathbf{x}(u,v)$ on a patch, the normal is *perpendicular* to the surface at $\mathbf{x}$. At $\mathbf{x}$, the surface has a *tangent plane*; it is defined

---

[3]This algorithm was described for cubics in Section 3.4 and for degree n Bézier curves in Section 4.3.

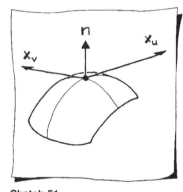

Sketch 51.
Tangent plane and normal.

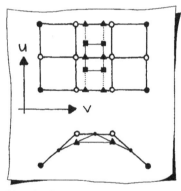

Sketch 52.
A schematic description of the 3-stage algorithm.

by $\mathbf{x}, \mathbf{x}_u, \mathbf{x}_v$—it is a point and two vectors. The normal $\mathbf{n}$ is a unit vector,[4] defined by

$$\mathbf{n} = \frac{\mathbf{x}_u \wedge \mathbf{x}_v}{\|\mathbf{x}_u \wedge \mathbf{x}_v\|}. \tag{6.18}$$

The geometry is explained in Sketch 51.

A note of caution: Since we are dividing by $\|\mathbf{x}_u \wedge \mathbf{x}_v\|$ in order to normalize, this term should not be zero! In some degenerate cases it can be, so a zero division check is usually a good idea. If this term is within tolerance of zero, you could use local geometry to make an informed guess. Returning the zero vector may or may not be an acceptable solution; depending on the application.

The main ingredients for $\mathbf{n}$ are $\mathbf{x}, \mathbf{x}_u$, and $\mathbf{x}_v$. The de Casteljau algorithm is the evaluation technique normally used for Bézier patches. As presented in Section 6.7, computing $\mathbf{x}_u$ and $\mathbf{x}_v$ would require two patch evaluations. However, there is a trick to compute these three ingredients almost simultaneously. Referring to the schematic diagram of a degree $m = 2 \times n = 3$ patch in Sketch 52, the algorithm proceeds as follows. First, compute $n - 1$ levels of the de Casteljau algorithm for all $m + 1$ rows of control points; these are all with respect to v, and depicted by triangles in the sketch. You now have two columns with $m + 1$ points each; compute $m - 1$ levels of the de Casteljau algorithm for each of them with parameter u. You are now left with four points, depicted by squares in the sketch, which form a bilinear patch. Its tangent plane at (u, v) agrees with the surface's tangent plane at (u, v). Thus, we evaluate and compute the partials of this bilinear patch at (u, v). Keep in mind that the partials of the bilinear patch must be scaled by a factor that reflects the degree of the original patch. Let's call this the *3-stage de Casteljau* evaluation method: a slightly modified version of the 2-stage de Casteljau plus a bilinear patch evaluation.

EXAMPLE 6.9

Let's revisit our patch from Example 6.4, and let's set $(u, v) = (0.5, 0.5)$. Two levels of the de Casteljau algorithm for the three rows of control points gives

---

[4]Its length is one.

$$\begin{bmatrix} 0 \\ 0 \\ 6 \end{bmatrix}$$

$$\begin{bmatrix} 3 \\ 0 \\ 0 \end{bmatrix} \quad \begin{bmatrix} 1.5 \\ 0 \\ 3 \end{bmatrix}$$

$$\begin{bmatrix} 6 \\ 0 \\ 0 \end{bmatrix} \quad \begin{bmatrix} 4.5 \\ 0 \\ 0 \end{bmatrix} \quad \begin{bmatrix} 3 \\ 0 \\ 1.5 \end{bmatrix}$$

$$\begin{bmatrix} 9 \\ 0 \\ 6 \end{bmatrix} \quad \begin{bmatrix} 7.5 \\ 0 \\ 3 \end{bmatrix} \quad \begin{bmatrix} 6 \\ 0 \\ 1.5 \end{bmatrix}$$

$$\begin{bmatrix} 0 \\ 3 \\ 3 \end{bmatrix}$$

$$\begin{bmatrix} 3 \\ 3 \\ 0 \end{bmatrix} \quad \begin{bmatrix} 1.5 \\ 3 \\ 1.5 \end{bmatrix}$$

$$\begin{bmatrix} 6 \\ 3 \\ 0 \end{bmatrix} \quad \begin{bmatrix} 4.5 \\ 3 \\ 0 \end{bmatrix} \quad \begin{bmatrix} 3 \\ 3 \\ 0.75 \end{bmatrix}$$

$$\begin{bmatrix} 9 \\ 3 \\ 0 \end{bmatrix} \quad \begin{bmatrix} 7.5 \\ 3 \\ 0 \end{bmatrix} \quad \begin{bmatrix} 6 \\ 3 \\ 0 \end{bmatrix}$$

$$\begin{bmatrix} 0 \\ 6 \\ 6 \end{bmatrix}$$

$$\begin{bmatrix} 3 \\ 6 \\ 0 \end{bmatrix} \quad \begin{bmatrix} 1.5 \\ 6 \\ 3 \end{bmatrix}$$

$$\begin{bmatrix} 6 \\ 6 \\ 0 \end{bmatrix} \quad \begin{bmatrix} 4.5 \\ 6 \\ 0 \end{bmatrix} \quad \begin{bmatrix} 3 \\ 6 \\ 1.5 \end{bmatrix}$$

$$\begin{bmatrix} 9 \\ 6 \\ 6 \end{bmatrix} \quad \begin{bmatrix} 7.5 \\ 6 \\ 3 \end{bmatrix} \quad \begin{bmatrix} 6 \\ 6 \\ 1.5 \end{bmatrix}$$

We now collect these results and perform one level of two more de Casteljau algorithms:

$$\begin{bmatrix} 3 \\ 0 \\ 1.5 \end{bmatrix}$$

$$\begin{bmatrix} 3 \\ 3 \\ 0.75 \end{bmatrix} \quad \begin{bmatrix} 3 \\ 1.5 \\ 1.125 \end{bmatrix}$$

$$\begin{bmatrix} 3 \\ 6 \\ 1.5 \end{bmatrix} \quad \begin{bmatrix} 3 \\ 4.5 \\ 1.125 \end{bmatrix}$$

$$\begin{bmatrix} 6 \\ 0 \\ 1.5 \end{bmatrix}$$

$$\begin{bmatrix} 6 \\ 3 \\ 0 \end{bmatrix} \quad \begin{bmatrix} 6 \\ 1.5 \\ 0.75 \end{bmatrix}$$

$$\begin{bmatrix} 6 \\ 6 \\ 1.5 \end{bmatrix} \quad \begin{bmatrix} 6 \\ 4.5 \\ 0.75 \end{bmatrix}$$

Thus, the bilinear patch is spanned by the four points

$$\begin{bmatrix} \begin{bmatrix} 3 \\ 1.5 \\ 1.125 \end{bmatrix} & \begin{bmatrix} 6 \\ 1.5 \\ 0.75 \end{bmatrix} \\ \begin{bmatrix} 3 \\ 4.5 \\ 1.125 \end{bmatrix} & \begin{bmatrix} 6 \\ 4.5 \\ 0.75 \end{bmatrix} \end{bmatrix}.$$

This bilinear patch has the same tangent plane as the original patch $\mathbf{x}$. To form $\mathbf{x}_u$, we simply use the 2-stage de Casteljau algorithm to find the u-partial of the bilinear patch, and then scale this result by the degree of $\mathbf{x}$ in the u-direction, $m = 2$:

$$\mathbf{x}_u = 2(\begin{bmatrix} 4.5 \\ 4.5 \\ 0.9375 \end{bmatrix} - \begin{bmatrix} 4.5 \\ 1.5 \\ 0.9375 \end{bmatrix}) = \begin{bmatrix} 0 \\ 6 \\ 0 \end{bmatrix}$$

(compare with Example 6.6!). The v-partial is found similarly:

$$\mathbf{x}_v = 3(\begin{bmatrix} 6 \\ 3 \\ 0.75 \end{bmatrix} - \begin{bmatrix} 3 \\ 3 \\ 1.125 \end{bmatrix}) = \begin{bmatrix} 9 \\ 0 \\ -1.125 \end{bmatrix}$$

(compare with Example 6.5!). Then,

$$\mathbf{x}_u \wedge \mathbf{x}_v = \begin{bmatrix} -6.75 \\ 0 \\ -54 \end{bmatrix}.$$

After normalization, we finally have

$$\mathbf{n} = \begin{bmatrix} -0.1240 \\ 0 \\ -0.9922 \end{bmatrix}.$$

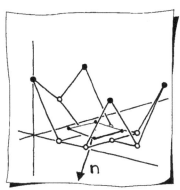

Sketch 53.
Computation of a normal vector.

The computations in Example (6.9) are illustrated in Sketch 53.

Of course, the role of the u and v directions could be switched in the 3-stage de Casteljau algorithm: Compute $m - 1$ levels of the de Casteljau algorithm for all $n + 1$ columns of control points; these are all with respect to u. You now have two rows with $n + 1$ points each; compute $n - 1$ levels of the de Casteljau algorithm for each of them with parameter v. You are now left with four points that define the same bilinear patch as above.

6.9 Changing Degrees

A Bézier patch has two degrees: m in the u-direction and n in the v-direction. Each of these degrees may be increased by a simple procedure. To be concrete, let's raise m to $m + 1$. The resulting control net—still describing the same surface—will have $n + 1$ columns of control points, each column containing $m + 2$ control points. These columns are simply obtained from the original columns by the process of degree elevation for curves as presented in Section 4.5.

EXAMPLE 6.10

Let us write the bilinear patch from Example 6.2 as a patch of degree 2 in u. We obtain the control net

$$\begin{bmatrix} \begin{bmatrix} 0 \\ 0 \\ 0 \end{bmatrix} & \begin{bmatrix} 0 \\ 1 \\ 0 \end{bmatrix} \\ \begin{bmatrix} 0.5 \\ 0.0 \\ 0 \end{bmatrix} & \begin{bmatrix} 0.5 \\ 1 \\ 0.5 \end{bmatrix} \\ \begin{bmatrix} 1 \\ 0 \\ 0 \end{bmatrix} & \begin{bmatrix} 1 \\ 1 \\ 1 \end{bmatrix} \end{bmatrix}.$$

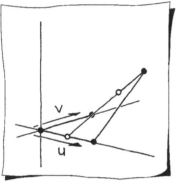

Sketch 54.
Degree elevation of a bilinear patch.

It is shown in Sketch 54.

To degree elevate in the v-direction, things follow the same pattern: Each of the $m + 1$ rows of the control net will be degree elevated to degree $n + 1$. Degree elevation in both the u- and v-directions is independent of the order in which it is done.

Degree reduction is also performed on a row-by-row or column-by-column basis, while repeatedly applying the curve algorithm. (See Section 4.6.)

6.10 Subdivision

Another curve operation was subdivision: This is the task of splitting one curve segment into two segments; see (4.10). Similarly, a patch may be subdivided into two patches. The u-parameter u_0 splits the domain unit square into two rectangles as shown in Figure 6.3. The patch is split along this isoparametric curve into two patches, together being identical to the original patch. The recipe for doing this cannot be simpler: Perform curve subdivision for each degree m column of the control net at parameter u_0.

EXAMPLE 6.11

Let's revisit our Example 6.4 patch and subdivide it at $u_0 = 0.5$. We have to subdivide each column of control points and arrive at the two new control nets.

$$
\begin{bmatrix} 0 \\ 0 \\ 6 \end{bmatrix} \qquad\qquad\qquad \begin{bmatrix} 3 \\ 0 \\ 0 \end{bmatrix}
$$

$$
\begin{bmatrix} 0 \\ 3 \\ 3 \end{bmatrix} \begin{bmatrix} 0 \\ 1.5 \\ 4.5 \end{bmatrix} \qquad\qquad \begin{bmatrix} 3 \\ 3 \\ 0 \end{bmatrix} \begin{bmatrix} 3 \\ 1.5 \\ 0 \end{bmatrix}
$$

$$
\begin{bmatrix} 0 \\ 6 \\ 6 \end{bmatrix} \begin{bmatrix} 0 \\ 4.5 \\ 4.5 \end{bmatrix} \begin{bmatrix} 0 \\ 3 \\ 4.5 \end{bmatrix} \quad \begin{bmatrix} 3 \\ 6 \\ 0 \end{bmatrix} \begin{bmatrix} 3 \\ 4.5 \\ 0 \end{bmatrix} \begin{bmatrix} 3 \\ 3 \\ 0 \end{bmatrix}
$$

$$
\begin{bmatrix} 6 \\ 0 \\ 0 \end{bmatrix} \qquad\qquad\qquad \begin{bmatrix} 9 \\ 0 \\ 6 \end{bmatrix}
$$

$$
\begin{bmatrix} 6 \\ 3 \\ 0 \end{bmatrix} \quad \begin{bmatrix} 6 \\ 1.5 \\ 0 \end{bmatrix} \qquad\qquad \begin{bmatrix} 9 \\ 3 \\ 0 \end{bmatrix} \quad \begin{bmatrix} 9 \\ 1.5 \\ 3 \end{bmatrix}
$$

$$
\begin{bmatrix} 6 \\ 6 \\ 0 \end{bmatrix} \quad \begin{bmatrix} 6 \\ 4.5 \\ 0 \end{bmatrix} \quad \begin{bmatrix} 6 \\ 3 \\ 0 \end{bmatrix} \qquad \begin{bmatrix} 9 \\ 6 \\ 6 \end{bmatrix} \quad \begin{bmatrix} 9 \\ 4.5 \\ 3 \end{bmatrix} \quad \begin{bmatrix} 9 \\ 3 \\ 3 \end{bmatrix}
$$

Gather the points along the diagonals of the schematic triangular diagrams of the de Casteljau algorithm to form the patch that lives over the domain $[0, 0.5] \times [0, 1]$ of the original patch:

$$
\begin{bmatrix}
\begin{bmatrix} 0 \\ 0 \\ 6 \end{bmatrix} & \begin{bmatrix} 3 \\ 0 \\ 0 \end{bmatrix} & \begin{bmatrix} 6 \\ 0 \\ 0 \end{bmatrix} & \begin{bmatrix} 9 \\ 0 \\ 6 \end{bmatrix} \\[1em]
\begin{bmatrix} 0 \\ 1.5 \\ 4.5 \end{bmatrix} & \begin{bmatrix} 3 \\ 1.5 \\ 0 \end{bmatrix} & \begin{bmatrix} 6 \\ 1.5 \\ 0 \end{bmatrix} & \begin{bmatrix} 9 \\ 1.5 \\ 3 \end{bmatrix} \\[1em]
\begin{bmatrix} 0 \\ 3 \\ 4.5 \end{bmatrix} & \begin{bmatrix} 3 \\ 3 \\ 0 \end{bmatrix} & \begin{bmatrix} 6 \\ 3 \\ 0 \end{bmatrix} & \begin{bmatrix} 9 \\ 3 \\ 3 \end{bmatrix}
\end{bmatrix} .
$$

Gather the points along the bases of the schematic triangular diagrams to form the patch that lives over $[0.5, 1.0] \times [0, 1]$ of the original patch.

$$
\begin{bmatrix}
\begin{bmatrix} 0 \\ 3 \\ 4.5 \end{bmatrix} & \begin{bmatrix} 3 \\ 3 \\ 0 \end{bmatrix} & \begin{bmatrix} 6 \\ 3 \\ 0 \end{bmatrix} & \begin{bmatrix} 9 \\ 3 \\ 3 \end{bmatrix} \\[1em]
\begin{bmatrix} 0 \\ 4.5 \\ 4.5 \end{bmatrix} & \begin{bmatrix} 3 \\ 4.5 \\ 0 \end{bmatrix} & \begin{bmatrix} 6 \\ 4.5 \\ 0 \end{bmatrix} & \begin{bmatrix} 9 \\ 4.5 \\ 3 \end{bmatrix} \\[1em]
\begin{bmatrix} 0 \\ 6 \\ 6 \end{bmatrix} & \begin{bmatrix} 3 \\ 6 \\ 0 \end{bmatrix} & \begin{bmatrix} 6 \\ 6 \\ 0 \end{bmatrix} & \begin{bmatrix} 9 \\ 6 \\ 6 \end{bmatrix}
\end{bmatrix} .
$$

This result is shown in Sketch 55.

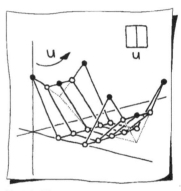

Sketch 55.
Subdivision of a surface.

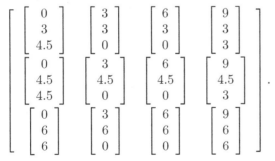

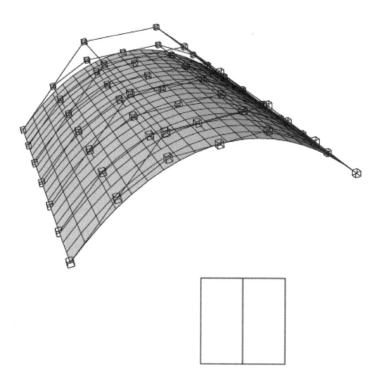

Figure 6.3.
Subdivision at $u = 0.5$ of a Bézier patch.

6.11   Ruled Bézier Patches

A ruled Bézier patch is linear in one isoparametric direction. As a result, this *ruled surface* may be written as

$$\mathbf{x}(u,v) = (1-v)\mathbf{x}(u,0) + v\mathbf{x}(u,1)$$

if the v-direction is linear, or

$$\mathbf{x}(u,v) = (1-u)\mathbf{x}(0,v) + u\mathbf{x}(1,v)$$

if the u-direction is linear. The bilinear surface of Section 6.2 is an example of a ruled surface.

Figure 6.4 illustrates that this type of surface supplies us with a simple method to fit a surface between two curves. The control points

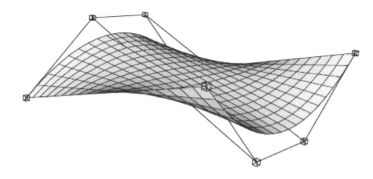

Figure 6.4.
A ruled Bézier patch.

of the ruled Bézier patch simply consist of the control points of the two given Bézier curves. In order to fit into the Bézier patch formalism, we must require that the two curves be of the same degree—this can always be achieved through degree elevation.

If the two curves corresponding to $v = 0$ and $v = 1$ are given by control points $\mathbf{b}_{0,0}, \ldots, \mathbf{b}_{m,0}$ and $\mathbf{b}_{0,1}, \ldots, \mathbf{b}_{m,1}$, then the ruled surface $\mathbf{x}(u,v)$ is given by

$$
\mathbf{x}(u,v) = [B_0^m(u), \ldots, B_m^m(u)] \begin{bmatrix} \mathbf{b}_{0,0} & \mathbf{b}_{0,1} \\ \vdots & \vdots \\ \mathbf{b}_{m,0} & \mathbf{b}_{m,1} \end{bmatrix} \begin{bmatrix} B_0^1(v) \\ B_1^1(v) \end{bmatrix} . \quad (6.19)
$$

A *developable surface* is a special ruled surface. This type of surface is important in manufacturing because it has the property that it can be "developed" by bending a piece of sheet metal, without tearing or stretching the metal. Special conditions exist for a ruled surface to be developable, namely that their Gaussian curvature (see Section 8.3) must be zero everywhere.

6.12 Functional Bézier Patches

Functional, or nonparametric Bézier patches are analogous to their curve counterparts from Section 4.8. The graph of a functional surface can be thought of as a parametric surface of the form:

$$\begin{bmatrix} x \\ y \\ z \end{bmatrix} = \begin{bmatrix} x(u) \\ y(v) \\ z(u,v) \end{bmatrix} = \begin{bmatrix} u \\ v \\ f(u,v) \end{bmatrix}.$$

Two dimensions are restricted to be linear polynomials, and the third is expressed as a function of them. This important feature makes functional Bézier patches *single-valued*. This makes these surfaces quite useful in some applications. For example, a surface designed for manufacturing via stamping sheet metal must be functional.

A functional Bézier patch defined over $[0,1] \times [0,1]$ is written as (6.8) with the entries in **B** taking the special values

$$\mathbf{b}_{i,j} = \begin{bmatrix} i/m \\ j/n \\ b_{i,j} \end{bmatrix}.$$

Examples for functional Bézier patches are given in Examples 6.4 and 6.10. Functional patches over an arbitrary rectangular region $[a,b] \times [c,d]$ are a direct generalization of functional Bézier curves over an arbitrary interval.

6.13   Monomial Patches

A monomial patch is defined as

$$\mathbf{x}(u,v) = \begin{bmatrix} 1 & u \ldots & u^m \end{bmatrix} \begin{bmatrix} \mathbf{a}_{0,0} & \ldots & \mathbf{a}_{0,n} \\ \vdots & & \vdots \\ \mathbf{a}_{m,0} & \ldots & \mathbf{a}_{m,n} \end{bmatrix} \begin{bmatrix} 1 \\ v \\ \vdots \\ v^n \end{bmatrix}, \quad (6.20)$$

or more concisely as

$$\mathbf{x}(u,v) = U^{\mathrm{T}} \mathbf{A} V.$$

Analogous to curves, the $\mathbf{a}_{0,0}$ element in the matrix represents a point on the patch at $(u,v) = (0,0)$. All other $\mathbf{a}_{i,j}$ are partial derivatives.

Conversion between curves in the monomial and the Bézier forms was examined in Section 3.7 and Section 4.4. Conversion between surfaces takes an analogous form:

$$\mathbf{a}_{i,j} = \binom{m}{i} \binom{n}{j} \Delta^{i,j} \mathbf{b}_{0,0}. \quad (6.21)$$

6.14 Exercises

1. Compute $\mathbf{x}(0.5, 1)$ for the patch of Example 6.4.

2. Compute $\mathbf{x}_{u,v}(0.5, 0.5)$ for the patch of Example 6.4 using the 3-stage de Casteljau evaluation method and the closed-form method of Section 6.6.

3. Using the patch of Example 6.4, elevate its degree from 2 to 3 in the u-direction.

4. Compute the normal to the patch in Example 6.2 at the three (u, v) pairs: $(0, 0), (0.5, 0.5), (1.0, 1.0)$.

5. Using the patch of Example 6.4, find the control nets after subdivision at $v = 0.5$. (Note that some of the calculation has been done in Example 6.9.)

6. There are two variations to the 2-stage de Casteljau evaluation method for Bézier surfaces: Create a u-isoparametric curve and evaluate at the v parameter, or create a v-isoparametric curve and evaluate at the u parameter. For a degree $m \times n$ patch, is the computation count the same for these two variations of the algorithm?

7. For a patch of degree $m \times n$, determine the computation count for the calculation of $\mathbf{x}_u$ and $\mathbf{x}_v$ via two 2-stage de Casteljau evaluations as in Section 6.7 versus the 3-stage approach in Section 6.8.

8. What are the control points for a biquadratic Bézier patch defining the function $\begin{bmatrix} x \\ y \\ x^2 \end{bmatrix}$ over the square $[0, 0] \times [1, 1]$? Then repeat for $[1, 1] \times [3, 3]$.

9. Convert the Bézier patch defined by the following control points to monomial form.

$$\begin{bmatrix} \begin{bmatrix} 0 \\ 0 \\ 0 \end{bmatrix} & \begin{bmatrix} 1 \\ 0 \\ 0 \end{bmatrix} & \begin{bmatrix} 2 \\ 0 \\ 0 \end{bmatrix} \\ \begin{bmatrix} 0 \\ 1 \\ 1 \end{bmatrix} & \begin{bmatrix} 1 \\ 1 \\ 1 \end{bmatrix} & \begin{bmatrix} 2 \\ 1 \\ 1 \end{bmatrix} \end{bmatrix}$$

10. What are the control points for the ruled Bézier patch which interpolates to the curves

$$\mathbf{c}(t) = \begin{bmatrix} 0 \\ 1 \\ 0 \end{bmatrix} B_0^2 + \begin{bmatrix} 1 \\ 1 \\ 1 \end{bmatrix} B_1^2 + \begin{bmatrix} 2 \\ 1 \\ 0 \end{bmatrix} B_2^2$$

and

$$\mathbf{d}(t) = \begin{bmatrix} -1 \\ 3 \\ -1 \end{bmatrix} B_0^3 + \begin{bmatrix} 0 \\ 3 \\ 1 \end{bmatrix} B_1^3 + \begin{bmatrix} 1 \\ 3 \\ 0 \end{bmatrix} B_2^3 + \begin{bmatrix} 2 \\ 3 \\ 0 \end{bmatrix} B_3^3.$$

11. The control points for the teapot from Figure 6.1 are on this book's web site. Get the data and display the teapot as a wire frame image.

Working with
Polynomial Patches

7

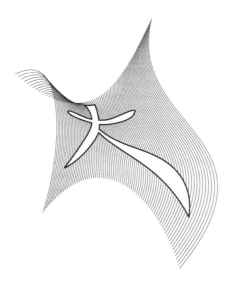

Figure 7.1.
A Bézier surface trimmed by a ConS. The 2D curve is illustrated in Figure 9.1.

So far, we have covered the basic surface theory. We will now learn to use surfaces for several applications.

7.1 Bicubic Interpolation

This is one of the simplest surface construction methods. Its uses are limited, but it prepares us for more general approaches. Suppose you are given 16 points $\mathbf{p}_{i,j}$ arranged as in Sketch 56 and also a pair of parameter values (u_i, v_j) with each of them.

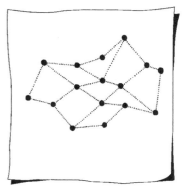

Sketch 56.
A 4×4 grid of points.

95

We seek an interpolating bicubic patch $\mathbf{x}(u, v)$, such that

$$
\begin{bmatrix}
\mathbf{p}_{0,0} & \mathbf{p}_{0,1} & \mathbf{p}_{0,2} & \mathbf{p}_{0,3} \\
\mathbf{p}_{1,0} & \mathbf{p}_{1,1} & \mathbf{p}_{1,2} & \mathbf{p}_{1,3} \\
\mathbf{p}_{2,0} & \mathbf{p}_{2,1} & \mathbf{p}_{2,2} & \mathbf{p}_{2,3} \\
\mathbf{p}_{3,0} & \mathbf{p}_{3,1} & \mathbf{p}_{3,2} & \mathbf{p}_{3,3}
\end{bmatrix}
=
\begin{bmatrix}
\mathbf{x}(u_0,v_0) & \mathbf{x}(u_0,v_1) & \mathbf{x}(u_0,v_2) & \mathbf{x}(u_0,v_3) \\
\mathbf{x}(u_1,v_0) & \mathbf{x}(u_1,v_1) & \mathbf{x}(u_1,v_2) & \mathbf{x}(u_1,v_3) \\
\mathbf{x}(u_2,v_0) & \mathbf{x}(u_2,v_1) & \mathbf{x}(u_2,v_2) & \mathbf{x}(u_2,v_3) \\
\mathbf{x}(u_3,v_0) & \mathbf{x}(u_3,v_1) & \mathbf{x}(u_3,v_2) & \mathbf{x}(u_3,v_3)
\end{bmatrix} . \tag{7.1}
$$

Following the matrix notation of Section 6.3, we may write each $\mathbf{x}(u_i, v_j)$ as a matrix product. We list an example:

$$
\mathbf{x}(u_1, v_2) =
$$

$$
\begin{bmatrix} B_0^3(u_1) & B_1^3(u_1) & B_2^3(u_1) & B_3^3(u_1) \end{bmatrix}
\begin{bmatrix}
\mathbf{b}_{0,0} & \mathbf{b}_{0,1} & \mathbf{b}_{0,2} & \mathbf{b}_{0,3} \\
\mathbf{b}_{1,0} & \mathbf{b}_{1,1} & \mathbf{b}_{1,2} & \mathbf{b}_{1,3} \\
\mathbf{b}_{2,0} & \mathbf{b}_{2,1} & \mathbf{b}_{2,2} & \mathbf{b}_{2,3} \\
\mathbf{b}_{3,0} & \mathbf{b}_{3,1} & \mathbf{b}_{3,2} & \mathbf{b}_{3,3}
\end{bmatrix}
\begin{bmatrix}
B_0^3(v_2) \\
B_1^3(v_2) \\
B_2^3(v_2) \\
B_3^3(v_2)
\end{bmatrix} . \tag{7.2}
$$

We may combine all individual equations for the $\mathbf{x}(u_i, v_j)$ into one matrix equation:

$$
\mathbf{P} = M^{\mathrm{T}} \mathbf{B} N \tag{7.3}
$$

where $\mathbf{P}$ is the given matrix of data points as in (7.1) and $\mathbf{B}$ is the matrix containing the unknown control points $\mathbf{b}_{i,j}$. The matrix M^{T} contains the values of the Bernstein polynomials at the given parameters:

$$
M^{\mathrm{T}} =
\begin{bmatrix}
B_0^3(u_0) & B_1^3(u_0) & B_2^3(u_0) & B_3^3(u_0) \\
B_0^3(u_1) & B_1^3(u_1) & B_2^3(u_1) & B_3^3(u_1) \\
B_0^3(u_2) & B_1^3(u_2) & B_2^3(u_2) & B_3^3(u_2) \\
B_0^3(u_3) & B_1^3(u_3) & B_2^3(u_3) & B_3^3(u_3)
\end{bmatrix} ,
$$

and N is given by

$$
N =
\begin{bmatrix}
B_0^3(v_0) & B_0^3(v_1) & B_0^3(v_2) & B_0^3(v_3) \\
B_1^3(v_0) & B_1^3(v_1) & B_1^3(v_2) & B_1^3(v_3) \\
B_2^3(v_0) & B_2^3(v_1) & B_2^3(v_2) & B_2^3(v_3) \\
B_3^3(v_0) & B_3^3(v_1) & B_3^3(v_2) & B_3^3(v_3)
\end{bmatrix} .
$$

Equation (7.3) is most conveniently decomposed into a sequence of linear systems. First, we define

$$
\mathbf{C} = M^{\mathrm{T}} \mathbf{B} \tag{7.4}
$$

which reduces (7.3) to

$$
\mathbf{P} = \mathbf{C} N .
$$

This constitutes four "vector" systems each with four linear equations, all having the same coefficient matrix N.[1] The unknowns are the rows of $\mathbf{C}$; the knowns are the rows of $\mathbf{P}$. For example, the second system of equations is given by

$$\begin{bmatrix} \mathbf{p}_{1,0} & \mathbf{p}_{1,1} & \mathbf{p}_{1,2} & \mathbf{p}_{1,3} \end{bmatrix} = \begin{bmatrix} \mathbf{c}_{1,0} & \mathbf{c}_{1,1} & \mathbf{c}_{1,2} & \mathbf{c}_{1,3} \end{bmatrix} N.$$

Having solved all four linear systems, we now tackle (7.4). This is also a set of four "vector" linear systems, this time with coefficient matrix M^{T}. The knowns are the columns of $\mathbf{C}$, the unknowns are the columns of $\mathbf{B}$. For example, the second system of equations is given by

$$\begin{bmatrix} \mathbf{c}_{0,1} \\ \mathbf{c}_{1,1} \\ \mathbf{c}_{2,1} \\ \mathbf{c}_{3,1} \end{bmatrix} = M^{\mathrm{T}} \begin{bmatrix} \mathbf{b}_{0,1} \\ \mathbf{b}_{1,1} \\ \mathbf{b}_{2,1} \\ \mathbf{b}_{3,1} \end{bmatrix}.$$

We solve all four linear systems and thus have found the desired matrix $\mathbf{B}$.

We could have solved (7.3) explicitly:

$$\mathbf{B} = (M^{\mathrm{T}})^{-1}\mathbf{P}N^{-1}, \tag{7.5}$$

but our approach—the *tensor product* approach—is much more efficient. This will become truly important for larger problems than the bicubic case.

EXAMPLE 7.1

Here is a simple example; the type you would do to debug your code. Let the given data be

$$\mathbf{p}_{i,j} = \begin{bmatrix} i \\ j \\ 0 \end{bmatrix} \qquad i = 0, 2, 3 \quad \text{and} \quad j = 0, 1, 2, 3$$

$$\mathbf{p}_{1,j} = \begin{bmatrix} 1 \\ j \\ 1 \end{bmatrix} \qquad j = 0, 1, 2, 3$$

and let the parameter pair for each $\mathbf{p}_{i,j}$ be $(i/3, j/3)$.

---

[1] Each vector system is comprised of a separate system for each coordinate.

First we solve for the rows of intermediate control points in $\mathbf{C}$ by solving the system

$$\begin{bmatrix} \mathbf{p}_{i,0} & \mathbf{p}_{i,1} & \mathbf{p}_{i,2} & \mathbf{p}_{i,3} \end{bmatrix} = \begin{bmatrix} \mathbf{c}_{i,0} & \mathbf{c}_{i,1} & \mathbf{c}_{i,2} & \mathbf{c}_{i,3} \end{bmatrix} \begin{bmatrix} 1 & 8/27 & 1/27 & 0 \\ 0 & 4/9 & 2/9 & 0 \\ 0 & 2/9 & 4/9 & 0 \\ 0 & 1/27 & 8/27 & 1 \end{bmatrix}$$

four times, for $i = 0, 1, 2, 3$. The columns of the matrix consist of each basis function evaluated at a particular v-parameter. Therefore, the columns must sum to one. When writing code, it is a good idea to check structures like this.

Draw your own sketch of the data. Notice that the $\mathbf{p}_{i,j}$ in each of the four curve interpolation problems lie on a line, and the points are uniformly spaced reflecting the parameter spacing. We know Bézier curves have linear precision (see Section 3.2). Therefore, we don't need to formally solve the system, as we know that the solution $\mathbf{C}$ is such that $\mathbf{c}_{i,j} = \mathbf{p}_{i,j}$.

Now, solve $\mathbf{C} = M^{\mathrm{T}}\mathbf{B}$. Refer to your sketch to realize that the four curve problems are pretty much identical. Isolate one of these, and sketch the curve interpolation problem. Using the linear precision property of Bézier curves again, we know the x and y coordinates of the $\mathbf{b}_{i,j}$ will be identical to those of the $\mathbf{c}_{i,j}$; just the z-coordinate needs computing. Since the parameters are identical in u and v, we have that $M^{\mathrm{T}} = N^{\mathrm{T}}$, and the four interpolation problems are

$$\begin{bmatrix} \mathbf{c}_{0,j} \\ \mathbf{c}_{1,j} \\ \mathbf{c}_{2,j} \\ \mathbf{c}_{3,j} \end{bmatrix} = M^{\mathrm{T}} \begin{bmatrix} \mathbf{b}_{0,j} \\ \mathbf{b}_{1,j} \\ \mathbf{b}_{2,j} \\ \mathbf{b}_{3,j} \end{bmatrix}$$

for $j = 0, 1, 2, 3$. Here is the solution for $j = 0$:

$$\begin{bmatrix} \begin{bmatrix} 0 \\ 0 \\ 0 \end{bmatrix} \\ \begin{bmatrix} 1 \\ 0 \\ 3 \end{bmatrix} \\ \begin{bmatrix} 2 \\ 0 \\ -1.5 \end{bmatrix} \\ \begin{bmatrix} 3 \\ 0 \\ 0 \end{bmatrix} \end{bmatrix}$$

Add these control points to your curve sketch and convince yourself that this indeed interpolates to the data.

We have not yet addressed the selection of the parameters u_i and v_j. In almost all instances of bicubic interpolation, these are set to $(u_0, u_1, u_2, u_3) = (0, 1/3, 2/3, 1)$ and analogously for the v_j. Theoretically, different values might improve the result, but finding those values does not seem worth the effort.

Let's revisit the concept of *tensor products* from Section 6.4. The interpolation problem above was solved, in a sense, by building the surface backwards. The tensor product property allows this to be possible. Sketch 57 illustrates the first stage of the interpolation process; the building of **C**. Each cubic curve in **C**, represented by control points depicted as squares, is built to pass through four of the $\mathbf{p}_{i,j}$, depicted as triangles. The next step in the interpolation process is illustrated in Sketch 58. Control points for the four cubic curves in **B**, depicted as circles, are found so that the cubics pass through the control points in **C**. The control points in **B** represent the control net for the Bézier patch passing through the $\mathbf{p}_{i,j}$. With this process in mind, take another look at the two-stage evaluation of a Bézier patch and its relation to the tensor product approach, as described in Section 6.4.

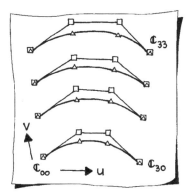

Sketch 57.
Step 1 in bicubic interpolation.

7.2 Interpolation using Higher Degrees

If we do not restrict the degrees m and n to be cubic, we still have the same structure as we did in the bicubic case. We are given an array of points

$$\mathbf{P} = \begin{bmatrix} \mathbf{p}_{0,0} & \cdots & \mathbf{p}_{0,n} \\ \vdots & & \vdots \\ \mathbf{p}_{m,0} & \cdots & \mathbf{p}_{m,n} \end{bmatrix}$$

with associated parameter values (u_i, v_j). We wish to find a Bézier patch

$$\mathbf{x}(u,v) = M^{\mathrm{T}} \mathbf{B} N$$

such that $\mathbf{x}(u_i, v_j) = \mathbf{p}_{i,j}$. In other words, we have to determine the matrix **B**, defined by

$$\mathbf{B} = \begin{bmatrix} \mathbf{b}_{0,0} & \cdots & \mathbf{b}_{0,n} \\ \vdots & & \vdots \\ \mathbf{b}_{m,0} & \cdots & \mathbf{b}_{m,n} \end{bmatrix}.$$

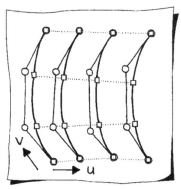

Sketch 58.
Step 2 in bicubic interpolation.

We follow the bicubic approach exactly: Write all interpolation conditions in matrix form:

$$\mathbf{P} = M^{\mathrm{T}}\mathbf{B}N$$

and define

$$\mathbf{C} = M^{\mathrm{T}}\mathbf{B}.$$

View

$$\mathbf{P} = \mathbf{C}N$$

as $m+1$ linear systems for the rows of $\mathbf{C}$, all having coefficient matrix N. Use the solution $\mathbf{C}$ as the knowns in

$$\mathbf{C} = M^{\mathrm{T}}\mathbf{B},$$

which is a set of $n+1$ linear systems for the columns of $\mathbf{B}$, all having the same coefficient matrix M^{T}.[2]

Two remarks: we could have used polynomials other than the Bernstein polynomials—for example, the monomials as per Section 6.13. We would still obtain the same interpolating surface.

We do warn against the use of polynomial interpolation using high degrees: the resulting surfaces tend to oscillate, just as in the curve case of Figure 5.2.

Sketch 59.
Four boundary curves of a surface.

7.3 Coons Patches

A common practical situation is this: Four boundary curves of a surface are designed, but the whole surface still has to be constructed. As Sketch 59 indicates, there are many possibilities.

The most widely used technique for this surface fitting problem goes back to S. Coons, who developed it in the 1960s for Ford. In the original development, the boundary curves could be arbitrary parametric curves, but with an eye on practical use, we restrict them to be Bézier curves.

We are then given the four boundary polygons of an array of points $\mathbf{b}_{i,j}; i = 0 \ldots m, j = 0 \ldots n$. A configuration with $m = n = 3$ looks like this:

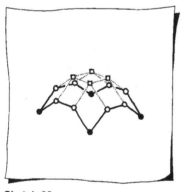

Sketch 60.
Four boundary polygons of a surface.

$$
\begin{array}{cccc}
\mathbf{b}_{0,0} & \mathbf{b}_{0,1} & \mathbf{b}_{0,2} & \mathbf{b}_{0,3} \\
\mathbf{b}_{1,0} & & & \mathbf{b}_{1,3} \\
\mathbf{b}_{2,0} & & & \mathbf{b}_{2,3} \\
\mathbf{b}_{3,0} & \mathbf{b}_{3,1} & \mathbf{b}_{3,2} & \mathbf{b}_{3,3}
\end{array}
$$

For this case, we would like to determine the missing four interior points, depicted by squares in Sketch 60.

---

[2]The matrix technique known as LU factorization should be used here for maximum efficiency.

Figure 7.2.

Completing a control net definition.

The general Coons formula is a blend of two linear interpolations and one bilinear interpolation:[3]

$$
\begin{aligned}
\mathbf{b}_{i,j} = {} & (1 - \frac{i}{m})\mathbf{b}_{0,j} + \frac{i}{m}\mathbf{b}_{m,j} \\
& + (1 - \frac{j}{n})\mathbf{b}_{i,0} + \frac{j}{n}\mathbf{b}_{i,n} \\
& - \begin{bmatrix} 1 - \frac{i}{m} & \frac{i}{m} \end{bmatrix} \begin{bmatrix} \mathbf{b}_{0,0} & \mathbf{b}_{0,n} \\ \mathbf{b}_{m,0} & \mathbf{b}_{m,n} \end{bmatrix} \begin{bmatrix} 1 - \frac{j}{n} \\ \frac{j}{n} \end{bmatrix} \\
& \text{for} \quad i = 1 \ldots m - 1 \quad \text{and} \quad j = 1 \ldots n - 1;
\end{aligned}
\tag{7.6}
$$

Figure 7.2 gives an example. The given boundary vertices are marked dark; the computed interior ones are shown in a lighter color.

EXAMPLE 7.2

For a numerical example, we reuse the control net from Example 6.4. We now treat the two interior control vertices $\mathbf{b}_{1,1}$ and $\mathbf{b}_{1,2}$ as unknowns. We then have

$$
\begin{bmatrix}
\begin{bmatrix} 0 \\ 0 \\ 6 \end{bmatrix} & \begin{bmatrix} 3 \\ 0 \\ 0 \end{bmatrix} & \begin{bmatrix} 6 \\ 0 \\ 0 \end{bmatrix} & \begin{bmatrix} 9 \\ 0 \\ 6 \end{bmatrix} \\
\begin{bmatrix} 0 \\ 3 \\ 3 \end{bmatrix} & \begin{bmatrix} ? \\ ? \\ ? \end{bmatrix} & \begin{bmatrix} ? \\ ? \\ ? \end{bmatrix} & \begin{bmatrix} 9 \\ 3 \\ 0 \end{bmatrix} \\
\begin{bmatrix} 0 \\ 6 \\ 6 \end{bmatrix} & \begin{bmatrix} 3 \\ 6 \\ 0 \end{bmatrix} & \begin{bmatrix} 6 \\ 6 \\ 0 \end{bmatrix} & \begin{bmatrix} 9 \\ 6 \\ 6 \end{bmatrix}
\end{bmatrix}.
$$

---

[3]See Section 6.2 for a review of bilinear interpolation.

It might help—in the hand computations—to note that the bilinear interpolation to the four corner points results in weighting factors

$$
\begin{bmatrix}
(1 - \frac{i}{m})(1 - \frac{j}{n}) & (1 - \frac{i}{m})\frac{j}{n} \\
\frac{i}{m}(1 - \frac{j}{n}) & \frac{i}{m}\frac{j}{n}
\end{bmatrix}
$$

Using the Coons formula for $m = 2$ and $n = 3$, we have

$$
\mathbf{b}_{1,1} = \frac{1}{2}\begin{bmatrix} 3 \\ 0 \\ 0 \end{bmatrix} + \frac{1}{2}\begin{bmatrix} 3 \\ 6 \\ 0 \end{bmatrix}
$$

$$
+ \frac{2}{3}\begin{bmatrix} 0 \\ 3 \\ 3 \end{bmatrix} + \frac{1}{3}\begin{bmatrix} 9 \\ 3 \\ 0 \end{bmatrix}
$$

$$
- \begin{bmatrix} 1/2 & 1/2 \end{bmatrix} \begin{bmatrix} \begin{bmatrix} 0 \\ 0 \\ 6 \\ 0 \\ 6 \\ 6 \end{bmatrix} & \begin{bmatrix} 9 \\ 0 \\ 6 \\ 9 \\ 6 \\ 6 \end{bmatrix} \end{bmatrix} \begin{bmatrix} 2/3 \\ 1/3 \end{bmatrix} = \begin{bmatrix} 3 \\ 3 \\ -4 \end{bmatrix}
$$

and

$$
\mathbf{b}_{1,2} = \frac{1}{2}\begin{bmatrix} 6 \\ 0 \\ 0 \end{bmatrix} + \frac{1}{2}\begin{bmatrix} 6 \\ 6 \\ 0 \end{bmatrix}
$$

$$
+ \frac{1}{3}\begin{bmatrix} 0 \\ 3 \\ 3 \end{bmatrix} + \frac{2}{3}\begin{bmatrix} 9 \\ 3 \\ 0 \end{bmatrix}
$$

$$
- \begin{bmatrix} 1/2 & 1/2 \end{bmatrix} \begin{bmatrix} \begin{bmatrix} 0 \\ 0 \\ 6 \\ 0 \\ 6 \\ 6 \end{bmatrix} & \begin{bmatrix} 9 \\ 0 \\ 6 \\ 9 \\ 6 \\ 6 \end{bmatrix} \end{bmatrix} \begin{bmatrix} 1/3 \\ 2/3 \end{bmatrix} = \begin{bmatrix} 6 \\ 3 \\ -5 \end{bmatrix}
$$

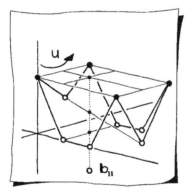

Sketch 61.

Three building blocks for a Coons patch.

Sketch 61 illustrates the construction of $\mathbf{b}_{1,1}$ in terms of the three components: two points generated from two linear interpolations, and one point generated by a bilinear interpolation.

7.4 Bicubic Hermite Interpolation

Hermite interpolation for curves used points and tangent vectors as input. In the surface case, we use points, partials, and mixed partials. They are usually grouped into a matrix of the form

$$\begin{bmatrix} \mathbf{x}(0,0) & \mathbf{x}_v(0,0) & \mathbf{x}_v(0,1) & \mathbf{x}(0,1) \\ \mathbf{x}_u(0,0) & \mathbf{x}_{uv}(0,0) & \mathbf{x}_{uv}(0,1) & \mathbf{x}_u(0,1) \\ \mathbf{x}_u(1,0) & \mathbf{x}_{uv}(1,0) & \mathbf{x}_{uv}(1,1) & \mathbf{x}_u(1,1) \\ \mathbf{x}(1,0) & \mathbf{x}_v(1,0) & \mathbf{x}_v(1,1) & \mathbf{x}(1,1) \end{bmatrix}. \tag{7.7}$$

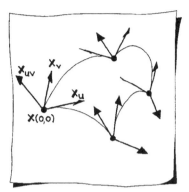

Sketch 62.
Input for bicubic Hermite interpolation.

Note how the coefficients in the matrix are grouped into four 2×2 partitions, each holding the data pertaining to one corner. The coefficients are shown in Sketch 62[4].

We are looking for a solution in Bézier from. Along the four patch boundaries, we are faced with four cubic Hermite curve interpolation problems. The resulting control vertices are thus

$$\mathbf{b}_{0,0} = \mathbf{x}(0,0) \qquad \mathbf{b}_{3,0} = \mathbf{x}(1,0)$$

$$\mathbf{b}_{0,1} = \mathbf{b}_{0,0} + \frac{1}{3}\mathbf{x}_v(0,0) \qquad \mathbf{b}_{3,1} = \mathbf{b}_{3,0} + \frac{1}{3}\mathbf{x}_v(1,0)$$

$$\mathbf{b}_{1,0} = \mathbf{b}_{0,0} + \frac{1}{3}\mathbf{x}_u(0,0) \qquad \mathbf{b}_{2,0} = \mathbf{b}_{3,0} - \frac{1}{3}\mathbf{x}_u(1,0)$$

$$\mathbf{b}_{0,3} = \mathbf{x}(0,1) \qquad \mathbf{b}_{3,3} = \mathbf{x}(1,1)$$

$$\mathbf{b}_{0,2} = \mathbf{b}_{0,3} - \frac{1}{3}\mathbf{x}_v(0,1) \qquad \mathbf{b}_{3,2} = \mathbf{b}_{3,3} - \frac{1}{3}\mathbf{x}_v(1,1)$$

$$\mathbf{b}_{1,3} = \mathbf{b}_{0,3} + \frac{1}{3}\mathbf{x}_u(0,1) \qquad \mathbf{b}_{2,3} = \mathbf{b}_{3,3} - \frac{1}{3}\mathbf{x}_u(1,1)$$

The remaining interior control vertices are found using the twist vector data. Recall the twist at each corner takes on a simple form, as given in (6.16) and (6.17). For example, at the $(0,0)$ corner:

$$\mathbf{x}_{uv}(0,0) = 9[\mathbf{b}_{1,1} - \mathbf{b}_{1,0} - \mathbf{b}_{0,1} + \mathbf{b}_{0,0}],$$

and we can solve for $\mathbf{b}_{1,1}$:

$$\mathbf{b}_{1,1} = \frac{1}{9}\mathbf{x}_{uv}(0,0) + \mathbf{b}_{0,1} + \mathbf{b}_{1,0} - \mathbf{b}_{0,0}.$$

---

[4]They are not drawn to the correct scale.

This same process at the other corners yields

$$\mathbf{b}_{2,1} = -\frac{1}{9}\mathbf{x}_{uv}(1,0) + \mathbf{b}_{3,1} - \mathbf{b}_{3,0} + \mathbf{b}_{2,0}$$

$$\mathbf{b}_{1,2} = -\frac{1}{9}\mathbf{x}_{uv}(0,1) + \mathbf{b}_{1,3} - \mathbf{b}_{0,3} + \mathbf{b}_{0,2}$$

$$\mathbf{b}_{2,2} = \frac{1}{9}\mathbf{x}_{uv}(1,1) - \mathbf{b}_{3,3} + \mathbf{b}_{2,3} + \mathbf{b}_{3,2}.$$

EXAMPLE 7.3

Suppose we are given the following "Hermite" matrix

$$\begin{bmatrix}
\begin{bmatrix} 0 \\ 0 \\ 0 \end{bmatrix} & \begin{bmatrix} 0 \\ 3 \\ 3 \end{bmatrix} & \begin{bmatrix} 0 \\ 3 \\ -3 \end{bmatrix} & \begin{bmatrix} 0 \\ 3 \\ 0 \end{bmatrix} \\
\begin{bmatrix} 3 \\ 0 \\ 3 \end{bmatrix} & \begin{bmatrix} 0 \\ 0 \\ 0 \end{bmatrix} & \begin{bmatrix} 0 \\ 0 \\ 0 \end{bmatrix} & \begin{bmatrix} 3 \\ 0 \\ 3 \end{bmatrix} \\
\begin{bmatrix} 3 \\ 0 \\ -3 \end{bmatrix} & \begin{bmatrix} 0 \\ 0 \\ 0 \end{bmatrix} & \begin{bmatrix} 0 \\ 0 \\ 0 \end{bmatrix} & \begin{bmatrix} 3 \\ 0 \\ -3 \end{bmatrix} \\
\begin{bmatrix} 3 \\ 0 \\ 0 \end{bmatrix} & \begin{bmatrix} 0 \\ 3 \\ 3 \end{bmatrix} & \begin{bmatrix} 0 \\ 3 \\ -3 \end{bmatrix} & \begin{bmatrix} 3 \\ 3 \\ 0 \end{bmatrix}
\end{bmatrix}$$

The Bézier patch for this data is illustrated in Figure 7.3. The Bézier points for this patch:

$$\begin{bmatrix}
\begin{bmatrix} 0 \\ 0 \\ 0 \end{bmatrix} & \begin{bmatrix} 0 \\ 1 \\ 1 \end{bmatrix} & \begin{bmatrix} 0 \\ 2 \\ 1 \end{bmatrix} & \begin{bmatrix} 0 \\ 3 \\ 0 \end{bmatrix} \\
\begin{bmatrix} 1 \\ 0 \\ 1 \end{bmatrix} & \begin{bmatrix} 1 \\ 1 \\ 2 \end{bmatrix} & \begin{bmatrix} 1 \\ 2 \\ 2 \end{bmatrix} & \begin{bmatrix} 1 \\ 3 \\ 1 \end{bmatrix} \\
\begin{bmatrix} 2 \\ 0 \\ 1 \end{bmatrix} & \begin{bmatrix} 2 \\ 1 \\ 2 \end{bmatrix} & \begin{bmatrix} 2 \\ 2 \\ 2 \end{bmatrix} & \begin{bmatrix} 2 \\ 3 \\ 1 \end{bmatrix} \\
\begin{bmatrix} 3 \\ 0 \\ 0 \end{bmatrix} & \begin{bmatrix} 3 \\ 1 \\ 1 \end{bmatrix} & \begin{bmatrix} 3 \\ 2 \\ 1 \end{bmatrix} & \begin{bmatrix} 3 \\ 3 \\ 0 \end{bmatrix}
\end{bmatrix}$$

Notice that all four twists are the zero vector. This is certainly a valid choice, but not recommended in practice because it will not, in general, produce patches with nice shape.

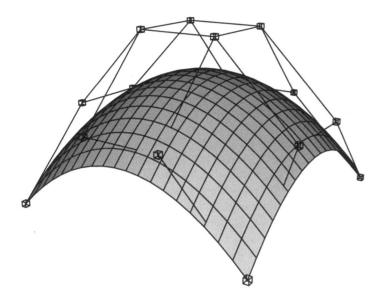

Figure 7.3.
A bicubic Bézier patch with zero twists.

Cubic Hermite interpolation is not used much any more in practice. Good, meaningful twist vectors are too hard to come by.[5] If one has the four boundary curves, then an easier way to define the whole patch is by using the Coons approach from Section 7.3.

7.5 Trimmed Patches

If we create any parametric curve $(u(t), v(t))$ in the domain of a surface $\mathbf{x}(u, v)$, it will be mapped to a curve on the surface, or ConS. This ConS is of the form $\mathbf{x}(u(t), v(t))$. Sketch 63 and Figure 7.1 illustrate this and the fact that the shape of the 3D ConS is a distorted version of the domain curve. ConS are mainly used for modification of tensor product surfaces by a technique known as "trimming." A trimmed surface has certain areas of it marked as "invalid" or "invisible" by a set of ConS. Sketch 64 gives an example. Here, two planes

Sketch 63.
A trimmed surface.

[5]There are a number of twist estimation techniques; see an advanced CAGD text.

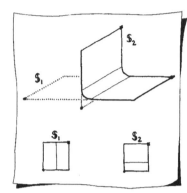

Sketch 64.
Two surfaces trimmed for a blend.

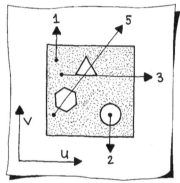

Sketch 65.
In/out test for domain curves.

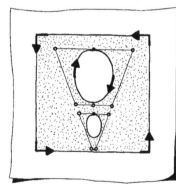

Sketch 66.
Orientation of trim curves.

are trimmed so that a blending surface can be placed between them. The dashed parts of the planes are considered "invisible," and would not be displayed.

If the domain curve is itself a Bézier curve of degree p and $\mathbf{x}(u,v)$ is degree $m \times n$, then the ConS will be, in general, of degree $(m+n)p$. Of course, an isoparametric line in the domain corresponds to an isoparametric curve on the surface, and this is a special ConS that you are familiar with already; it will be of degree m or n. Also, recall the ConS from Example 6.2.

If the domain curve of a ConS is *closed*, then it divides the domain into two parts: points inside the curve and points outside. In the same way, the closed ConS divides the surface into two parts. If we want to know, for an arbitrary point (u,v) in the domain, if it lies inside the domain curve, take an arbitrary ray emanating from (u,v). Then, count the number of its intersections with all domain curves and the boundary of the surface domain. If it is even, (u,v) is outside, it is inside otherwise. See Sketch 65 for an illustration. For programming purposes, there are no "arbitrary" rays. Rays parallel to the u- or v-direction will typically suffice.[6] Tangencies count as two intersections! It is a common practice to orient the trim curves, as illustrated in Sketch 66, so that the inside trim curves are clockwise and the outer-boundary is counterclockwise.

Trimmed surfaces are a bread-and-butter tool in all CAD/CAM systems. They can arise in many applications, the most common one resulting from the intersection between two surfaces. The resulting intersection curve is a ConS on either of the two surfaces.

ConS can cause numerical problems—so wherever they are used, special care is needed. For example, their storage is an issue. It is wiser to simply store the domain curve, not the 3D curve. Otherwise, when transformations are applied to the surface, round-off could cause the 3D curve to leave the surface.

7.6 Least Squares Approximation

In many cases, data points do not come on a rectangular grid which aligns with the patch boundaries. If the data points come from a laser digitizer, for example, we cannot expect them to have any recognizable structure whatsoever, except that they are numbered. We are thus dealing with a set of points $\mathbf{p}_k$, with k ranging from 0 to some

---

[6]If the domain curves have a tendency to follow the isoparametric lines, then this choice can make the computation sort of tricky.

number $K-1$—this number will typically be in the hundreds or thousands. We also assume that each data point $\mathbf{p}_k$ has a corresponding pair of parameters (u_k, v_k). See Figure 7.4 for an illustration.

We wish to find a Bézier patch that fits the data as good as possible. Let it be given by a control net with coefficients $\mathbf{b}_{i,j}$, with $i = 0, \ldots, m$ and $j = 0, \ldots, n$. Before we formulate our problem, we need to introduce a new notation.

Instead of writing a Bézier patch as a matrix product (6.7), we introduce a *linearized* notation. We will number all control points

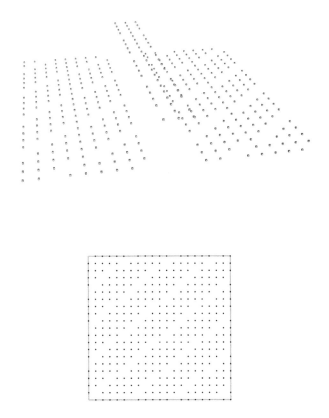

Figure 7.4.
Data for least squares surface approximation. Each 3D data point is accompanied
by a parameter pair. Domain lower left corresponds to range back right.

linearly, simply counting them as we traverse the control net row by row. For example, we may write a bilinear patch as

$$\mathbf{x}(u, v) =$$

$$\begin{bmatrix} B_0^1(u)B_0^1(v), & B_0^1(u)B_1^1(v), & B_1^1(u)B_0^1(v), & B_1^1(u)B_1^1(v) \end{bmatrix} \begin{bmatrix} \mathbf{b}_{0,0} \\ \mathbf{b}_{0,1} \\ \mathbf{b}_{1,0} \\ \mathbf{b}_{1,1} \end{bmatrix}.$$

This equation has, for the general case, $(m+1)(n+1)$ terms. Using this linear ordering, our patch equation may be written as

$$\mathbf{x}(u, v) = \begin{bmatrix} B_0^m(u)B_0^n(v), \ldots, B_m^m(u)B_n^n(v) \end{bmatrix} \begin{bmatrix} \mathbf{b}_{0,0} \\ \vdots \\ \mathbf{b}_{m,n} \end{bmatrix}.$$

Returning to our problem, we would like to achieve that each data point lies on the approximating surface. For the kth data point $\mathbf{p}_k$, this becomes $\mathbf{p}_k = \mathbf{x}(u_k, v_k)$ or

$$\mathbf{p}_k = \begin{bmatrix} B_0^m(u_k)B_0^n(v_k), \ldots, B_m^m(u_k)B_n^n(v_k) \end{bmatrix} \begin{bmatrix} \mathbf{b}_{0,0} \\ \vdots \\ \mathbf{b}_{m,n} \end{bmatrix}.$$

Combining all K of these equations, we obtain

$$\begin{bmatrix} \mathbf{p}_0 \\ \vdots \\ \vdots \\ \vdots \\ \mathbf{p}_{K-1} \end{bmatrix} = \begin{bmatrix} B_0^m(u_0)B_0^n(v_0) & \cdots & B_m^m(u_0)B_n^n(v_0) \\ & \vdots & \\ & \vdots & \\ & \vdots & \\ B_0^m(u_{K-1})B_0^n(v_{K-1}) & \cdots & B_m^m(u_{K-1})B_n^n(v_{K-1}) \end{bmatrix} \begin{bmatrix} \mathbf{b}_{0,0} \\ \vdots \\ \mathbf{b}_{m,n} \end{bmatrix}.$$

We abbreviate this to

$$\mathbf{P} = M\mathbf{B}. \tag{7.8}$$

These are K equations in $(m+1)(n+1)$ unknowns. For the example of the bicubic case, we have 16 unknowns, but typically several hundred data points—thus the linear system (7.8) is *overdetermined*. It will in general not have an exact solution, but a good approximation is found by forming

$$M^{\mathrm{T}}\mathbf{P} = M^{\mathrm{T}}M\mathbf{B}, \tag{7.9}$$

which is the same procedure that solved the analogous curve problem, see Section 5.4. The system in (7.9) is called the system of *normal equations*. Our linear system has a square coefficient matrix $M^{\mathrm{T}}M$, with $(m+1)(n+1)$ rows and columns. In the bicubic case, we would thus have to solve a linear system with 16 equations in 16 unknowns.[7] The solution $\mathbf{B}$ is the *least squares approximation* in Bézier form to the given data. The term "least squares" refers to the fact that this solution minimizes the sum of the squared distances of each data point to the resulting surface.

EXAMPLE 7.4

We give a very simple example for $m = n = 1$ and $K = 5$. Let the data points be

$$\mathbf{p}_0 = \begin{bmatrix} -2 \\ -2 \\ 1 \end{bmatrix}, \quad \mathbf{p}_1 = \begin{bmatrix} 2 \\ 2 \\ 1 \end{bmatrix}, \quad \mathbf{p}_2 = \begin{bmatrix} 1 \\ 0 \\ 0 \end{bmatrix}, \quad \mathbf{p}_3 = \begin{bmatrix} 0 \\ 1 \\ 0 \end{bmatrix}, \quad \mathbf{p}_4 = \begin{bmatrix} 0.5 \\ 0.5 \\ 1 \end{bmatrix}.$$

Let their parameter values be

$$\begin{aligned}
(u_0, v_0) &= (0, 0) \\
(u_1, v_1) &= (1, 1) \\
(u_2, v_2) &= (1, 0) \\
(u_3, v_3) &= (0, 1) \\
(u_4, v_4) &= (0.5, 0.5).
\end{aligned}$$

---

[7] Just as for the curves, these are vector equations. By using the LU-factorization, we decompose the matrix $M^{\mathrm{T}}M$ once and then apply this to each coordinate.

The overdetermined linear system (7.8) is given by

$$
\begin{bmatrix}
\begin{bmatrix} -2 \\ -2 \\ 1 \end{bmatrix} \\
\begin{bmatrix} 2 \\ 2 \\ 1 \end{bmatrix} \\
\begin{bmatrix} 1 \\ 0 \\ 0 \end{bmatrix} \\
\begin{bmatrix} 0 \\ 1 \\ 0 \end{bmatrix} \\
\begin{bmatrix} 0.5 \\ 0.5 \\ 1 \end{bmatrix}
\end{bmatrix}
=
\begin{bmatrix}
1 & 0 & 0 & 0 \\
0 & 0 & 0 & 1 \\
0 & 0 & 1 & 0 \\
0 & 1 & 0 & 0 \\
0.25 & 0.25 & 0.25 & 0.25
\end{bmatrix}
\begin{bmatrix}
\mathbf{b}_{0,0} \\
\mathbf{b}_{0,1} \\
\mathbf{b}_{1,0} \\
\mathbf{b}_{1,1}
\end{bmatrix}.
$$

Multiplying both sides by the transpose of the coefficient matrix yields

$$
\begin{bmatrix}
\begin{bmatrix} -1.875 \\ -1.875 \\ 1 \end{bmatrix} \\
\begin{bmatrix} 0.125 \\ 1.125 \\ 0 \end{bmatrix} \\
\begin{bmatrix} 1.125 \\ 0.125 \\ 0 \end{bmatrix} \\
\begin{bmatrix} 2.125 \\ 2.125 \\ 1 \end{bmatrix}
\end{bmatrix}
=
\begin{bmatrix}
1+x & x & x & x \\
x & 1+x & x & x \\
x & x & 1+x & x \\
x & x & x & 1+x
\end{bmatrix}
\begin{bmatrix}
\mathbf{b}_{0,0} \\
\mathbf{b}_{0,1} \\
\mathbf{b}_{1,0} \\
\mathbf{b}_{1,1}
\end{bmatrix},
$$

where $x = 1/16$.

Solving the linear system (actually one each for x, y, z) yields

$$
\mathbf{b}_{0,0} = \begin{bmatrix} -1.95 \\ -1.95 \\ 0.9 \end{bmatrix}, \quad
\mathbf{b}_{0,1} = \begin{bmatrix} 0.05 \\ 1.05 \\ -0.1 \end{bmatrix}, \quad
\mathbf{b}_{1,0} = \begin{bmatrix} 1.05 \\ 0.05 \\ -0.1 \end{bmatrix}, \quad
\mathbf{b}_{1,1} = \begin{bmatrix} 2.05 \\ 2.05 \\ 0.9 \end{bmatrix}.
$$

This surface does not go through any of the data points exactly, but it is reasonably close to them. Figure 7.5 illustrates a bicubic least squares approximation example.

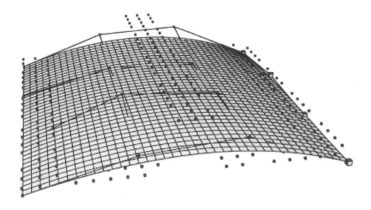

Figure 7.5.
Least squares approximation to the data in Figure 7.4.

Notice that if the number of data points K is equal to the number of control points $(m + 1)(n + 1)$, then we have interpolation. The system in (7.8) may be solved directly; there is no need to form the normal equations (7.9). However, the interpolation method from Section 7.2 is more stable since it guarantees a better distribution of parameter pairs.

One question remains: How do we obtain the parameter values (u_k, v_k) for the data point $\mathbf{p}_k$? This problem is hard to solve in general. But often, the data points can be projected into a plane; let's assume they can be projected into the (x, y)-plane for simplicity, as illustrated in Sketch 67. Each $\mathbf{p}_k$ is projected by simply dropping its z-coordinate, leaving a pair (x_k, y_k). We scale the (x_k, y_k) so that they fit into the unit square. Then we can simply set $u_k = x_k$ and $v_k = y_k$.

If the data cannot be projected into a plane, then the next step would be to look for a basic surface with a known parametrization which mimics the shape of the data. For example, a cylinder or a sphere might be a good choice. Sketch 68 illustrates how a cylinder would be used. Each data point would be projected onto a cylinder, thus generating a (θ, z) parameter pair. Just as above, scale these parameters to live in the unit square.

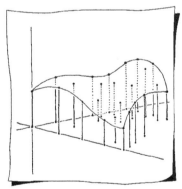

Sketch 67.
Projecting data points into the (x, y)-plane.

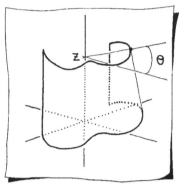

Sketch 68.
Projecting data points onto a cylinder.

7.7 Exercises

1. Outline the steps to solve the bicubic interpolation problem in Example 7.1 by constructing intermediate control points which represent curves with constant u.

2. If you have a system solver: Modify the data from Example 7.1 to use u-parameters $(0, 1/2, 2/3, 1)$. Compare the resulting surface to the one in the example.

3. Find the bilinear interpolant in monomial form to the data

$$\mathbf{p}_{0,0} = \begin{bmatrix} 0 \\ 0 \\ 0 \end{bmatrix} \quad \mathbf{p}_{1,0} = \begin{bmatrix} 1 \\ 0 \\ 0 \end{bmatrix} \quad \mathbf{p}_{0,1} = \begin{bmatrix} 0 \\ 2 \\ 1 \end{bmatrix} \quad \mathbf{p}_{1,1} = \begin{bmatrix} 1 \\ 2 \\ -1 \end{bmatrix}$$

with parameters $u_i = (0, 1)$ and $v_i = (0, 1)$.

4. Let an incomplete control net be given by

$$\begin{bmatrix} \begin{bmatrix} -1 \\ -1 \\ 1 \end{bmatrix} & \begin{bmatrix} 0 \\ -0.5 \\ 0.5 \end{bmatrix} & \begin{bmatrix} 1 \\ 0 \\ 0 \end{bmatrix} \\ \begin{bmatrix} -0.5 \\ 0 \\ 0.5 \end{bmatrix} & \begin{bmatrix} ? \\ ? \\ ? \end{bmatrix} & \begin{bmatrix} 1.5 \\ 1 \\ 0.5 \end{bmatrix} \\ \begin{bmatrix} 0 \\ 1 \\ 0 \end{bmatrix} & \begin{bmatrix} 1 \\ 1.5 \\ 0.5 \end{bmatrix} & \begin{bmatrix} 2 \\ 2 \\ 1 \end{bmatrix} \end{bmatrix}.$$

Find the missing control point using the Coons technique.

5. The Hermite form of a surface $\mathbf{x}$ is defined by the elements in the matrix (7.7) and the Hermite basis functions from Section 5.6. Give an expression for the surface in matrix form.

6. Give a rough computation count as to why the tensor product approach is more efficient for bicubic interpolation than the explicit solution in (7.5).

7. Suppose you are designing a library of Bézier modification routines. You need to have a trim surface function which will allow the user to trim along whole isoparametric curves. How can you achieve

this special trim functionality without actually implementing trim surfaces into your library?

8. Change the fifth data point from Example 7.6 to

$$\mathbf{p}_4 = \begin{bmatrix} 0.5 \\ 0.5 \\ 0.5 \end{bmatrix}$$

and resolve the problem.

Shape

8

Figure 8.1.
A common tool to investigate surface geometry is that of reflection lines. A pattern of reflection lines on a B-spline surface is shown. Figure courtesy of H. Theisel.

When the first author worked in the CAD/CAM department of Mercedes-Benz, Germany, he often had to communicate with designers who traditionally have little training in mathematics. A designer thinks in terms such as "fair," "smooth," or "sweet." How can such concepts be incorporated into computer programs? As it turned out, the central concept of any kind of shape description is *curvature*, and, luckily, it lends itself to a very intuitive understanding.

8.1   The Frenet Frame

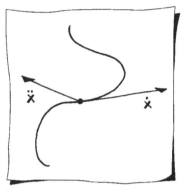

Sketch 69.
Two derivative vectors at a point on a curve.

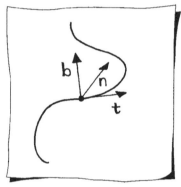

Sketch 70.
The Frenet frame.

We will discuss the shape of a curve in *local* terms, i.e., we will talk about the curve's shape at a particular point $\mathbf{x}(t)$. In order to do this, it would be helpful to have a local coordinate system at $\mathbf{x}(t)$, thus enabling us to express local curve properties in terms of this system. We will base the construction of such a system on the first and second derivatives of the curve: $\dot{\mathbf{x}}(t)$ and $\ddot{\mathbf{x}}(t)$, illustrated in Sketch 69.

Neither of these two derivatives is (in general) of unit length, nor are they orthogonal to each other. Yet we may use them to define our desired coordinate system using normalization and cross products. Our coordinate system will have origin $\mathbf{x}(t)$ and three axes $\mathbf{t}, \mathbf{b}, \mathbf{n}$ which we define as

$$\mathbf{t} = \frac{\dot{\mathbf{x}}(t)}{\|\dot{\mathbf{x}}(t)\|}, \tag{8.1}$$

$$\mathbf{b} = \frac{\dot{\mathbf{x}}(t) \wedge \ddot{\mathbf{x}}(t)}{\|\dot{\mathbf{x}}(t) \wedge \ddot{\mathbf{x}}(t)\|}, \tag{8.2}$$

$$\mathbf{n} = \mathbf{b} \wedge \mathbf{t}. \tag{8.3}$$

These three vectors form a local coordinate system at $\mathbf{x}(t)$ and are called the *Frenet frame* at $\mathbf{x}(t)$. The vectors $\mathbf{t}, \mathbf{b}$, and $\mathbf{n}$ are called unit tangent, binormal, and normal vector, respectively. While vectors have no fixed position in space, it is customary to think of them as emanating from $\mathbf{x}(t)$. For an illustration, see Sketch 70. In Section 8.2 we will see why $\mathbf{n}$ was chosen as above instead of $\mathbf{t} \wedge \mathbf{b}$.

A note of caution: if either $\dot{\mathbf{x}}(t)$ or $\dot{\mathbf{x}}(t) \wedge \ddot{\mathbf{x}}(t)$ are the zero vector, then the Frenet frame is not defined. Such points do not occur very often, and can usually be handled by suitable special-case considerations.

For planar curves in the (x, y)-plane, the binormal vector $\mathbf{b}(t)$ is constant. This is true for any plane curve, as shown in Sketch 71.

As we let $\mathbf{x}(t)$ trace out points on our curve, the corresponding Frenet frames will also slide along the curve, always being an orthonormal system, but changing orientation. A use for this is positioning objects along a curve. In Figure 8.2, the object is the letter F enclosed in a box. For several points on the planar curve, the Frenet frame is computed. The lower left hand corner of the box corresponds to the point on the curve, and the bottom and left side correspond to $\mathbf{t}$ and $\mathbf{n}$, respectively. This ensures that the letter always at the same location relative to the Frenet frame.

Figure 8.2.
A letter being moved along a curve.

EXAMPLE 8.1

Let a planar cubic Bézier curve be given by the control points

$$\begin{bmatrix} 0 \\ 0 \end{bmatrix}, \begin{bmatrix} 0 \\ 1 \end{bmatrix}, \begin{bmatrix} 1 \\ 1 \end{bmatrix}, \begin{bmatrix} 2 \\ 1 \end{bmatrix}.$$

Let's compute the Frenet frame at $t = 0$ and $t = 1$. First, we must compute the first and second derivatives at the ends:

$$\dot{\mathbf{x}}(0) = \begin{bmatrix} 0 \\ 3 \end{bmatrix}, \qquad \dot{\mathbf{x}}(1) = \begin{bmatrix} 3 \\ 0 \end{bmatrix},$$

$$\ddot{\mathbf{x}}(0) = \begin{bmatrix} 6 \\ -6 \end{bmatrix}, \qquad \ddot{\mathbf{x}}(1) = \begin{bmatrix} 0 \\ 0 \end{bmatrix}.$$

This makes the Frenet frames at the ends:

$$\mathbf{t}(0) = \begin{bmatrix} 0 \\ 1 \\ 0 \end{bmatrix}, \qquad \mathbf{t}(1) = \begin{bmatrix} 1 \\ 0 \\ 0 \end{bmatrix},$$

$$\mathbf{b}(0) = \begin{bmatrix} 0 \\ 0 \\ -1 \end{bmatrix}, \qquad \mathbf{b}(1) = \begin{bmatrix} 0 \\ 0 \\ 0 \end{bmatrix},$$

$$\mathbf{n}(0) = \begin{bmatrix} 1 \\ 0 \\ 0 \end{bmatrix}, \qquad \mathbf{n}(1) = \begin{bmatrix} 0 \\ 0 \\ 0 \end{bmatrix}.$$

This curve and the Frenet frame at $t = 0$ is illustrated in Sketch 72. The Frenet frame at $t = 1$ is undefined.

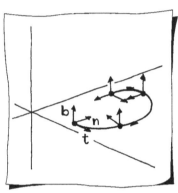

Sketch 71.
Planar curves have a constant binormal vector.

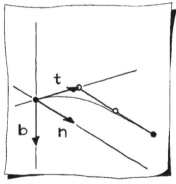

Sketch 72.
A Frenet frame at one point on a Bézier curve.

8.2 Curvature and Torsion

How is the Frenet frame related to the shape of a curve? As we move along the curve, we observe how the frame changes. The more the curve is bent, the faster the frame will change. We say that the rate of change of the unit tangent vector $\mathbf{t}$ denotes the *curvature* of the curve. For a straight line, the curvature is zero; for a circle, it is constant. A formula for the curvature, denoted by κ, is given by

$$\kappa(t) = \frac{\|\dot{\mathbf{x}}(t) \wedge \ddot{\mathbf{x}}(t)\|}{\|\dot{\mathbf{x}}(t)\|^3} \tag{8.4}$$

The curvature is also related to the circle which best approximates the curve at $\mathbf{x}(t)$. This circle, called the *osculating circle*, has radius $\rho = 1/\kappa$, and its center is

$$\mathbf{c}(t) = \mathbf{x}(t) + \rho(t)\mathbf{n}(t). \tag{8.5}$$

The osculating circle's center is in the direction of the normal vector $\mathbf{n}$. The osculating circle lies in the *osculating plane*, which is spanned by $\mathbf{t}$ and $\mathbf{n}$.

EXAMPLE 8.2

Let's continue with the curve in Example 8.1 by computing the curvature at $t = 0$ and $t = 1$: we have

$$\kappa(0) = \frac{2}{3} \quad \text{and} \quad \kappa(1) = 0.$$

This agrees nicely with our intuitive notion of "curvedness": at the left endpoint $(t = 0)$, the curve bends sharply; at the right endpoint $(t = 1)$, it is flat. Sketch 73 illustrates the center of the osculating circle at $t = 0$:

$$\mathbf{c}(0) = \left[\begin{array}{c} 3/2 \\ 0 \end{array} \right].$$

The center at $t = 1$ is undefined since $\rho = 1/0$.

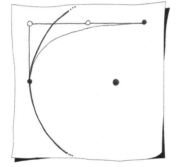

Sketch 73.
Curvature of a Bézier curve.

For the special case of Bézier curves, curvature may be computed without having to compute derivatives explicitly. At $t = 0$, the curvature of a Bézier curve is given by

$$\kappa(0) = 2\frac{n-1}{n} \frac{\text{area}[\mathbf{b}_0, \mathbf{b}_1, \mathbf{b}_2]}{\|\mathbf{b}_1 - \mathbf{b}_0\|^3}. \tag{8.6}$$

We see that the curve has zero curvature if the three points $\mathbf{b}_0, \mathbf{b}_1, \mathbf{b}_2$ are collinear. Similarly, at $t = 1$, the curvature is given by

$$\kappa(1) = 2\frac{n-1}{n}\frac{\text{area}[\mathbf{b}_{n-2}, \mathbf{b}_{n-1}, \mathbf{b}_n]}{\|\mathbf{b}_n - \mathbf{b}_{n-1}\|^3}. \tag{8.7}$$

EXAMPLE 8.3

Let's return to the Bézier curve from Example 8.1 and compute $\kappa(0)$ using (8.6). We have (using (1.4) for the area of a triangle)

$$\text{area}[\mathbf{b}_0, \mathbf{b}_1, \mathbf{b}_2] = \frac{1}{2}|\det\begin{bmatrix} 1 & 1 & 1 \\ 0 & 0 & 1 \\ 0 & 1 & 1 \end{bmatrix}| = \frac{1}{2}.$$

Thus (taking into account $\|\mathbf{b}_1 - \mathbf{b}_0\| = 1$)

$$\kappa(0) = 2 \cdot \frac{2}{3} \cdot \frac{1}{2} = \frac{2}{3}.$$

This is indeed the same as in Example 8.2.

If the curvature is desired at parameter values other than 0 or 1, the process of subdivision can be employed: Just subdivide at the desired parameter value and proceed as above.

A 3D curve has, by definition, nonnegative curvature. For 2D curves, we may assign a sign to the curvature by defining

$$\kappa(t) = \frac{\det\begin{bmatrix} \dot{\mathbf{x}}(t), & \ddot{\mathbf{x}}(t) \end{bmatrix}}{\|\dot{\mathbf{x}}(t)\|^3}. \tag{8.8}$$

With this notion of curvature, we may define *inflection points*: These are points where the curvature changes sign. For Bézier curves, signed curvature is easily introduced by noting that the quantity $\text{area}[\mathbf{b}_0, \mathbf{b}_1, \mathbf{b}_2]$ can be assigned a sign in 2D. This sign does not actually belong to the curvature, but it gives an indication of how the curve is changing in relation to the right-hand rule.

EXAMPLE 8.4

Let's take the curve from Example 8.1 and change it to create an inflection point by setting

$$\mathbf{b}_3 = \begin{bmatrix} 1 \\ 2 \end{bmatrix}.$$

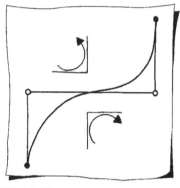

Sketch 74.
A planar cubic Bézier curve with an inflection point.

See Sketch 74 for an illustration.

The signed curvature at $t = 0$ is computed as in Example 8.3, however we no longer take the absolute value of the determinant. Thus, $\kappa(0) = -2/3$.

At $t = 1$:

$$\kappa(1) = \frac{4}{3}\frac{1}{2} \det \begin{bmatrix} 1 & 1 & 1 \\ 0 & 1 & 1 \\ 1 & 1 & 2 \end{bmatrix} = \frac{2}{3}$$

The curvatures at the endpoints are the same in value, but opposite in sign. Since this is a cubic polynomial, the curvature is continuous along the curve. This implies that the curvature must be zero somewhere.

The curvature of a curve is the most significant descriptor of its shape. Most commercial systems allow a user to check the shape of a curve by displaying its *curvature plot*. This is simply the graph of $\kappa(t)$.

The curvature plot is a very sensitive instrument for judging a curve's shape. Wherever the curve is not "perfect," the curvature plot will exhibit oscillations. Figures 8.3 and 8.4 illustrate. Each figure shows a B-spline curve[1] and also its curvature plot. While both curves can hardly be distinguished by just looking at their graphs, the curvature plots tell the two curves apart immediately. The underlying curve is a side view of a Mercedes-Benz curve, taken from the hood. The first author introduced the concept of curvature plots to Mercedes-Benz designers—it soon became their tool of choice for judging the shape of curves.

The *torsion* measures the change in a curve's binormal vector. A formula for the torsion, denoted by τ, is given by

$$\tau(t) = \frac{\det[\dot{\mathbf{x}}, \ddot{\mathbf{x}}, \dddot{\mathbf{x}}]}{\|\dot{\mathbf{x}} \wedge \ddot{\mathbf{x}}\|^2}.$$

Recall from Section 8.1 that the binormal of a planar curve is constant, therefore, a quadratic curve has zero torsion. For the special case of Bézier curves, torsion, just as curvature, may be computed without having to compute derivatives explicitly. At $t = 0$, the torsion of a Bézier curve is given by

$$\tau(0) = \frac{3}{2}\frac{n-2}{n}\frac{\text{volume}[\mathbf{b}_0, \mathbf{b}_1, \mathbf{b}_2, \mathbf{b}_3]}{\text{area}[\mathbf{b}_0, \mathbf{b}_1, \mathbf{b}_2]^2}.$$

[1]See Chapter 10.

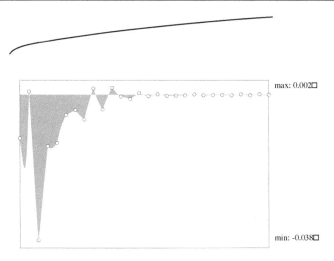

Figure 8.3.
A curve with its curvature plot.

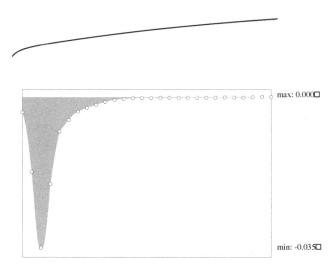

Figure 8.4.
An improved curve with its curvature plot.

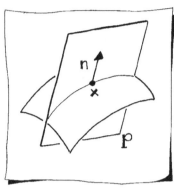

Sketch 75.
A normal section.

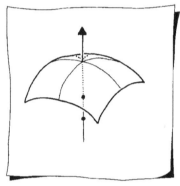

Sketch 76.
An elliptic point.

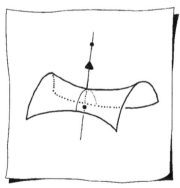

Sketch 77.
A saddle point.

8.3   Surface Curvatures

Defining the shape of surfaces is quite a bit harder than it was for curves.

We will need one basic tool: the concept of *normal curvature*. Let $\mathbf{x}(u, v)$ be a point on a surface and let $\mathbf{n}(u, v)$ be its normal. Any plane $\mathbf{P}$ through $\mathbf{x}$ which contains $\mathbf{n}$ will intersect the surface in a curve, see Sketch 75. This curve is called the *normal section* of $\mathbf{x}$ with respect to $\mathbf{P}$. It is planar by definition; we can compute its signed curvature at $\mathbf{x}$. This curvature $\kappa_{\mathbf{P}}$ is the normal curvature of the surface at point $\mathbf{x}$ with respect to the plane $\mathbf{P}$.

Now imagine rotating $\mathbf{P}$ around $\mathbf{n}$. For each new position of $\mathbf{P}$, we will get a new normal section, and hence a new normal curvature. Of all the normal curvatures at $\mathbf{x}$, one will be the largest, called $\kappa_{\max}$, and one will be the smallest, called $\kappa_{\min}$. These two curvatures are called the *principal curvatures* at $\mathbf{x}$. Depending on the sign of the principal curvatures, we may distinguish three cases:

1. Both $\kappa_{\min}$ and $\kappa_{\max}$ are positive or both are negative. Then $\mathbf{x}$ is called an *elliptic point* of the surface. Sketch 76 illustrates the center of the osculating circle (8.5) for each extreme curvature. All points on a sphere or on an ellipsoid are elliptic.

2. $\kappa_{\min}$ and $\kappa_{\max}$ are of opposite sign. Then $\mathbf{x}$ is called a *hyperbolic point* of the surface. Sketch 77 illustrates. Another term is *saddle point*. All points on hyperboloids and bilinear patches are hyperbolic.[2]

3. One of the principal curvatures is zero. Then $\mathbf{x}$ is called a *parabolic point*. Sketch 78 illustrates. Cylinders or cones are examples for this type.

These three cases are succinctly described by one quantity, namely the product K of $\kappa_{\min}$ and $\kappa_{\max}$:

$$K = \kappa_{\min}\kappa_{\max},$$

called *Gaussian curvature*. The sign of K determines which of the three cases best describes the shape of the surface near the point $\mathbf{x}$.

---

[2]The best "real life" example of surfaces with hyperbolic points are potato chips.

The Gaussian curvature can be computed using the first and second derivatives of the surface. We define

$$F = \det \begin{bmatrix} \mathbf{x}_u\mathbf{x}_u & \mathbf{x}_u\mathbf{x}_v \\ \mathbf{x}_u\mathbf{x}_v & \mathbf{x}_v\mathbf{x}_v \end{bmatrix}$$

and

$$S = \det \begin{bmatrix} \mathbf{n}\mathbf{x}_{u,u} & \mathbf{n}\mathbf{x}_{u,v} \\ \mathbf{n}\mathbf{x}_{u,v} & \mathbf{n}\mathbf{x}_{v,v} \end{bmatrix}.$$

All quantities involved in F and S are easily computed. The two determinants are called first and second fundamental matrices of the surface at $\mathbf{x}$.

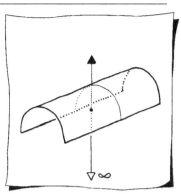

Sketch 78.
A parabolic point.

The Gaussian curvature is then given by

$$K = \frac{S}{F}. \tag{8.9}$$

Let's revisit the list from above:

1. An elliptic point corresponds to $K > 0$.

2. A hyperbolic point corresponds to $K < 0$.

3. A parabolic point corresponds to $K = 0$.[3]

Of course most surfaces are not composed entirely of one type of Gaussian curvature.

EXAMPLE 8.5

Let a bilinear surface be given by the four Bézier points

$$\mathbf{b}_{0,0} = \begin{bmatrix} 0 \\ 0 \\ 0 \end{bmatrix}, \quad \mathbf{b}_{1,0} = \begin{bmatrix} 1 \\ 0 \\ 0 \end{bmatrix}, \quad \mathbf{b}_{0,1} = \begin{bmatrix} 0 \\ 1 \\ 0 \end{bmatrix}, \quad \mathbf{b}_{1,1} = \begin{bmatrix} 1 \\ 1 \\ 1 \end{bmatrix}.$$

At $\mathbf{x}(0,0)$, we have

$$\mathbf{x}_u = \begin{bmatrix} 1 \\ 0 \\ 0 \end{bmatrix}, \mathbf{x}_v = \begin{bmatrix} 0 \\ 1 \\ 0 \end{bmatrix}, \mathbf{x}_{u,v} = \begin{bmatrix} 0 \\ 0 \\ 1 \end{bmatrix}, \mathbf{n} = \begin{bmatrix} 0 \\ 0 \\ 1 \end{bmatrix},$$

and the two second partials $\mathbf{x}_{uu}$ and $\mathbf{x}_{vv}$ are the zero vector. Therefore,

$$F = \det \begin{bmatrix} 1 & 0 \\ 0 & 1 \end{bmatrix} = 1$$

---

[3]There are more surfaces with vanishing K: These are called *developable surfaces*.

and

$$S = \det \begin{bmatrix} 0 & 1 \\ 1 & 0 \end{bmatrix} = -1.$$

and hence

$$K = -1,$$

which is negative as expected.

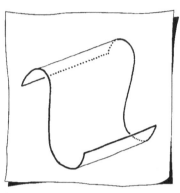

Sketch 79.
A surface with zero Gaussian curvature.

While Gaussian curvature is an important tool, it is not a panacea: the surface in Sketch 79 is, intuitively, quite curved. Yet its Gaussian curvature vanishes for every point on it since $\kappa_{\min} = 0$ everywhere.

More shape measures exist; we note two of them. The first one is *mean curvature M*. It is defined by

$$M = \frac{1}{2}[\kappa_{\min} + \kappa_{\max}].$$

It can easily be computed like this:

$$M = \frac{[\mathbf{n}\mathbf{x}_{vv}]\mathbf{x}_u^2 - 2[\mathbf{n}\mathbf{x}_{uv}][\mathbf{x}_u\mathbf{x}_v] + [\mathbf{n}\mathbf{x}_{uu}]\mathbf{x}_v^2}{F}.$$

The mean curvature is zero for surfaces that are called *minimal*. Such surfaces resemble the shape of soap bubbles.

The second curvature that is of practical use is *absolute curvature A*. It is given by

$$A = |\kappa_{\min}| + |\kappa_{\max}|$$

and measures the curvature of a surface in the most reliable way from an intuitive viewpoint.

A similar expression, sometimes called the RMS (root mean square) curvature is defined as

$$R = \sqrt{\kappa_{\min}^2 + \kappa_{\max}^2}.$$

It can be computed as

$$R = \sqrt{4M^2 - 2K}.$$

A somewhat unusual application of this curvature is shown in Figure 8.5. There, a native Indian vessel was digitized (left) and fitted with a B-spline surface. The RMS curvatures of this surface were then used as a texture map on the surface, thus "painting" it with its own curvature.

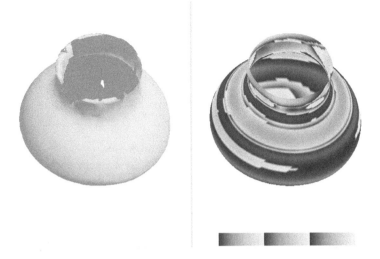

Figure 8.5.
A digitized vessel (left) and the RMS curvatures of a B-spline approximation (right).
Figure courtesy of M-S. Bae.

8.4   Reflection Lines

The surface curvatures which we encountered above are not neces-
sarily intuitive to designers trying to create "beautiful" shapes.[4] A
different surface tool is used more often. It is based on the simulation
of an automotive design studio.

Once a car prototype has been built, it is placed in a studio with
a ceiling filled with parallel fluorescent light bulbs. Their reflections
in the car's surface give designers crucial feedback on the quality of
their product: "Flowing" reflection patterns are good, "wiggly" ones
are bad.

Nowadays, these light patterns can be simulated well before a prototype is built. Simple surface geometry code can keep a company
from building an expensive prototype with flawed shape.

The principle is straightforward: When plotting a surface, highlight
areas in which reflections will occur. The simplest model to do this
is as follows. For any point **x** on the surface, compute its normal **n**.
Let **L** denote a line light source; see Sketch 80.

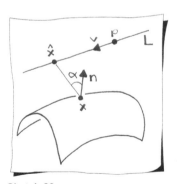

Sketch 80.
A reflection line model.

---

[4]For the design of curves, the tool of a curvature plot is well-accepted.

If the angle α between $\mathbf{n}$ and $\mathbf{L}$ is small, the normal points to the light source $\mathbf{L}$, and the corresponding region of the surface is highlighted. If $\alpha \neq 0$ (within a user-defined tolerance), then no highlighting happens.

To compute α, we first find the point $\hat{\mathbf{x}}$ on $\mathbf{L}$ which is closest to $\mathbf{x}$. If $\mathbf{L}$ is given by a point $\mathbf{p}$ and a vector $\mathbf{v}$, then $\hat{\mathbf{x}}$ is given by

$$\hat{\mathbf{x}} = \mathbf{p} + \frac{\mathbf{v}[\mathbf{x} - \mathbf{p}]}{\|\mathbf{v}\|^2}\mathbf{v}. \tag{8.10}$$

Now α is given by the angle between $\mathbf{n}$ and $\hat{\mathbf{x}} - \mathbf{x}$.

The above method defines a curve on the given surface which is determined by the light line $\mathbf{L}$; it is sometimes referred to as an *isophote*. Figures 8.6 and 8.7 illustrates this method. The B-spline surface in Figure 8.6 has some shape imperfections which are not too obvious from the shaded image. The reflection line display in Figure 8.7 reveals them clearly. After a smoothing algorithm was applied, the improved pattern of Figure 8.1 emerged.

More complicated models also include the eye position of the observer. Different designers prefer different simulations of the design studio. Any of these can be computed using the tools from this chapter.

But keep in mind how this tool differs from the curvatures above: It depends on extraneous information, such as light lines and eye points. A surface may look good for some set of light lines, but bad

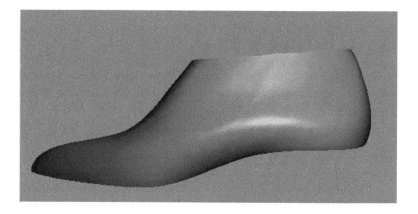

Figure 8.6.

A B-spline model of a shoe last. Figure courtesy of H. Theisel.

Figure 8.7.
Reflection lines: The center part of the previous figure with a set of reflection lines.
Figure courtesy of H. Theisel.

for another. The curvatures are an absolute measure, independent of extraneous artifacts.

8.5 Exercises

1. What is the Frenet frame at $t = 0$ and $t = 1$ for the quadratic Bézier curve with control points

$$\mathbf{b}_0 = \begin{bmatrix} 0 \\ 0 \\ 0 \end{bmatrix}, \quad \mathbf{b}_1 = \begin{bmatrix} 1 \\ 1 \\ 0 \end{bmatrix}, \quad \mathbf{b}_2 = \begin{bmatrix} 2 \\ 0 \\ 1 \end{bmatrix}.$$

2. Let a cubic Bézier curve be given by

$$\begin{bmatrix} 0 \\ 1 \\ 0 \end{bmatrix}, \begin{bmatrix} 0 \\ 0 \\ 0 \end{bmatrix}, \begin{bmatrix} -1 \\ 0 \\ 0 \end{bmatrix}, \begin{bmatrix} -1 \\ 1 \\ 1 \end{bmatrix}.$$

Find $\kappa(0)$ and $\kappa(1)$.

3. Let a bilinear surface be given by four Bézier points $\mathbf{b}_{i,j}$

$$\mathbf{b}_{0,0} = \begin{bmatrix} 0 \\ 0 \\ 0 \end{bmatrix}, \quad \mathbf{b}_{1,0} = \begin{bmatrix} 1 \\ 0 \\ 0 \end{bmatrix}, \quad \mathbf{b}_{0,1} = \begin{bmatrix} 0 \\ 0 \\ 1 \end{bmatrix}, \quad \mathbf{b}_{1,1} = \begin{bmatrix} 1 \\ 0 \\ 1 \end{bmatrix}.$$

   Find the Gaussian curvature at $\mathbf{x}(0,0)$.

4. For the bilinear patch above, find the mean curvature M and the RMS curvature R at $\mathbf{x}(0,0)$.

5. Suppose a light line is defined by

$$\mathbf{p} = \begin{bmatrix} 0 \\ 0 \end{bmatrix} \quad \text{and} \quad \mathbf{v} = \begin{bmatrix} 1 \\ 1 \end{bmatrix}.$$

   Find the closest point $\hat{\mathbf{x}}$ on the line to the given point $\mathbf{x} = \begin{bmatrix} 2 \\ -1 \end{bmatrix}$.
   Be sure to sketch the problem and the solution.

6. Repeat problems 3 and 4 after all control points are scaled by a factor of 2.

Composite Curves 9

Figure 9.1.
"Big" in Chinese; designed with piecewise Bézier curves.

While Bézier curves are a powerful tool, they are not very convenient when it comes to modeling complex curves. For these, *composite* curves are typically used—also known as *splines*. A composite curve is composed of pieces, and thus the term *piecewise* curve is also used.

In this chapter, we will examine Bézier curves that are linked together, and we will outline the conditions for various kinds of smoothness at the transition from one curve segment to the next.

9.1 Piecewise Bézier Curves

Local and global parameters are discussed in Section 4.7. Composite Bézier curves provide an application of these concepts. As illustrated in Sketch 81, each Bézier curve is defined over an interval. The intervals are defined by two successive *knots* $[u_i, u_{i+1}]$. All knots are referred to as the *knot sequence*. We will use the nota-

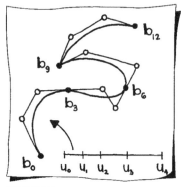

Sketch 81.
Four cubic Bézier curves linked
together.

tion $\Delta_i = u_{i+1} - u_i$. Piecewise curves defined over a common knot sequence are called *spline curves*.

When examining a property of a spline curve $\mathbf{s}$, we will refer to the global parameter u within the knot vector. When examining a property of the ith Bézier curve $\mathbf{s}_i$, defined over $[u_i, u_{i+1}]$, we will refer to its local parameter $t \in [0, 1]$. Recall from Section 2.1 that

$$t = \frac{u - u_i}{\Delta_i}.$$

We will also use the term *junction point*. These are curve segment end points:

$$\mathbf{s}(u_i) = \mathbf{s}_i(0) = \mathbf{s}_{i-1}(1).$$

Now let's look at the interplay between the local and global parameters when we consider derivatives of a Bézier curve. The derivative of a spline curve at u, must incorporate the derivative of the ith curve segment $\mathbf{s}_i(t)$ when $u \in [u_i, u_{i+1}]$, and it must also incorporate the length of the interval in order to put the derivative in the context of $\mathbf{s}(u)$. In other words, the chain rule must be invoked and it leads to

$$\frac{\mathrm{d}\mathbf{s}(u)}{\mathrm{d}u} = \frac{\mathrm{d}\mathbf{s}_i(t)}{\mathrm{d}t} \frac{\mathrm{d}t}{\mathrm{d}u}$$
$$= \frac{1}{\Delta_i} \frac{\mathrm{d}\mathbf{s}_i(t)}{\mathrm{d}t}.$$

The focus here will be Bézier curves, and in particular, at the junction points. In order to examine the smoothness of two curves at a junction point, we will need the derivatives of the Bézier curve at their endpoints. For the junction u_1 between the first two cubic Bézier curves in Sketch 81, this corresponds to

$$\dot{\mathbf{s}}_0(1) = \frac{3}{\Delta_0}\Delta\mathbf{b}_2 \quad \text{and} \quad \dot{\mathbf{s}}_1(0) = \frac{3}{\Delta_1}\Delta\mathbf{b}_3.$$

The second derivatives follow similarly:

$$\ddot{\mathbf{s}}_0(1) = \frac{6}{\Delta_0^2}\Delta^2\mathbf{b}_1 \quad \text{and} \quad \ddot{\mathbf{s}}_1(0) = \frac{6}{\Delta_1^2}\Delta^2\mathbf{b}_3.$$

9.2 C^1 and G^1 Continuity

Let us consider two cubic Bézier curves, one defined over the domain interval $[u_0, u_1]$, the other one over $[u_1, u_2]$. Let the Bézier points of the "left" curve be $\mathbf{b}_0, \mathbf{b}_1, \mathbf{b}_2, \mathbf{b}_3$ and let those of the "right" one be

$\mathbf{b}_3, \mathbf{b}_4, \mathbf{b}_5, \mathbf{b}_6$. These two curve segments form one global curve which is continuous.

What conditions must be met such that the two segments form a *differentiable* or C^1 curve over the interval $[u_0, u_2]$? Taking derivatives of each segment at parameter value u_1 and equating these derivatives, we arrive at the condition

$$\mathbf{b}_3 = \frac{\Delta_1}{\Delta}\mathbf{b}_2 + \frac{\Delta_0}{\Delta}\mathbf{b}_4 \qquad (9.1)$$

where $\Delta = u_2 - u_0$.

This may be interpreted geometrically: The three points $\mathbf{b}_2, \mathbf{b}_3, \mathbf{b}_4$ must be collinear and they must be an affine map of the 1D parameter values u_0, u_1, u_2. This is illustrated in Sketch 82. Note also that (9.1) may be expressed as

$$\text{ratio}(\mathbf{b}_2, \mathbf{b}_3, \mathbf{b}_4) = \frac{\Delta_0}{\Delta_1}.$$

For the special case of $u_0, u_1, u_2 = 0, 1, 2$, we get

$$\mathbf{b}_3 = \frac{1}{2}\mathbf{b}_2 + \frac{1}{2}\mathbf{b}_4.$$

EXAMPLE 9.1

Let two cubics be given by Bézier points

$$\mathbf{b}_0, \mathbf{b}_1, \mathbf{b}_2, \mathbf{b}_3 = \begin{bmatrix} 1 \\ 0 \end{bmatrix}, \begin{bmatrix} 0 \\ 1 \end{bmatrix}, \begin{bmatrix} 1 \\ 2 \end{bmatrix}, \begin{bmatrix} 2 \\ 2 \end{bmatrix}$$

and

$$\mathbf{b}_3, \mathbf{b}_4, \mathbf{b}_5, \mathbf{b}_6 = \begin{bmatrix} 2 \\ 2 \end{bmatrix}, \begin{bmatrix} 4 \\ 2 \end{bmatrix}, \begin{bmatrix} 6 \\ -2 \end{bmatrix}, \begin{bmatrix} 3 \\ -2 \end{bmatrix}.$$

Let the first curve be defined over $[u_0, u_1] = [0, 1]$ and the second one over $[u_1, u_2] = [1, 3]$. The C^1 condition requires

$$\mathbf{b}_3 = \frac{2}{3}\mathbf{b}_2 + \frac{1}{3}\mathbf{b}_4 = \begin{bmatrix} 2 \\ 2 \end{bmatrix}.$$

Our two segments satisfy this condition, and hence we have a C^1 curve. Sketch 83 illustrates.

But if we set $[u_1, u_2] = [1, 2]$, we do not have a C^1 curve any more since $\mathbf{b}_3$ is not the midpoint of $\mathbf{b}_2$ and $\mathbf{b}_4$!

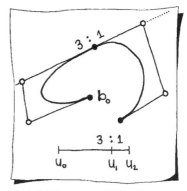

Sketch 82.
Two C^1 cubics and their domains.

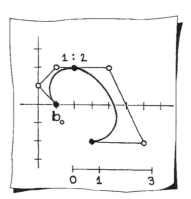

Sketch 83.
A C^1 piecewise cubic curve.

The last part of Example 9.1 raises an important point: Two cubics may be C^1 with one selection of parameter intervals, but not with respect to another. This is best understood by interpreting the parameter interval $[u_0, u_2]$ as a time interval. As time progresses from u_0 to u_2, the corresponding point on the curve moves from $\mathbf{b}_0$ to $\mathbf{b}_6$. If this motion is C^1, then the point's velocity must change continuously. This implies it must travel faster over "long" parameter intervals, and slower over "short" ones.

The concept of C^1 involves the speed of traversal of the curve, not just the curve's geometry. This speed is controlled by the spacing of the parameter intervals. If we are interested purely in the shape of the curve, the parameters should thus not be consulted at all.

From a purely geometric viewpoint, a curve is smooth or G^1 continuous (for Geometrically continuous) if its tangent line varies continuously. For two cubic Bézier curves, this simply implies that $\mathbf{b}_2, \mathbf{b}_3$, and $\mathbf{b}_4$ must be collinear.

9.3 C^2 and G^2 Continuity

We again consider two cubic Bézier segments with the notation of Section 9.2. Assuming the two segments form one C^1 curve, comparing second derivatives at parameter value u_1, we get the following C^2 condition:

$$-\frac{\Delta_1}{\Delta_0}\mathbf{b}_1 + \frac{\Delta}{\Delta_0}\mathbf{b}_2 = \frac{\Delta}{\Delta_1}\mathbf{b}_4 - \frac{\Delta_0}{\Delta_1}\mathbf{b}_5. \tag{9.2}$$

This has a simple geometric interpretation, see Sketch 85. Each side of (9.2) defines a point. The left hand side, using the "left" Bézier segment, defines $\mathbf{d}_-$ and the right hand side, using the "right" one, defines $\mathbf{d}_+$:

$$\mathbf{d}_- = -\frac{\Delta_1}{\Delta_0}\mathbf{b}_1 + \frac{\Delta}{\Delta_0}\mathbf{b}_2 \qquad \mathbf{d}_+ = \frac{\Delta}{\Delta_1}\mathbf{b}_4 - \frac{\Delta_0}{\Delta_1}\mathbf{b}_5. \tag{9.3}$$

The C^2 condition requires that $\mathbf{d}_- = \mathbf{d}_+$, which will be called $\mathbf{d}$.

Suppose that the curves are C^2, then rearranging (9.3) gives

$$\mathbf{b}_2 = \frac{\Delta_1}{\Delta}\mathbf{b}_1 + \frac{\Delta_0}{\Delta}\mathbf{d}, \qquad \mathbf{b}_4 = \frac{\Delta_1}{\Delta}\mathbf{d} + \frac{\Delta_0}{\Delta}\mathbf{b}_5,$$

and finally

$$\text{ratio}(\mathbf{b}_1, \mathbf{b}_2, \mathbf{d}) = \text{ratio}(\mathbf{d}, \mathbf{b}_4, \mathbf{b}_5) = \frac{\Delta_0}{\Delta_1}.$$

This geometry is illustrated in Sketch 84.

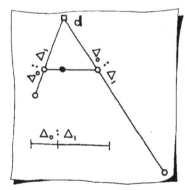

Sketch 84.
The C^2 condition.

EXAMPLE 9.2

We revisit the two Bézier segments from Example 9.1. Do they form
a C^2 curve? We have to compute the point $\mathbf{d}$ in two ways. From the
"left" curve, we get

$$\mathbf{d}_- = -2\mathbf{b}_1 + 3\mathbf{b}_2 = \begin{bmatrix} 3 \\ 4 \end{bmatrix}$$

From the "right" curve, we get

$$\mathbf{d}_+ = \frac{3}{2}\mathbf{b}_4 - \frac{1}{2}\mathbf{b}_5 = \begin{bmatrix} 3 \\ 4 \end{bmatrix}.$$

Since both values for $\mathbf{d}$ agree, the curve is indeed C^2. See Sketch 85
for the geometry.

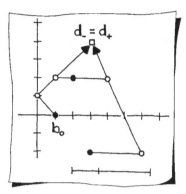

Sketch 85.
A C^2 curve.

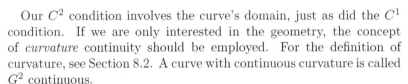

Our C^2 condition involves the curve's domain, just as did the C^1
condition. If we are only interested in the geometry, the concept
of *curvature* continuity should be employed. For the definition of
curvature, see Section 8.2. A curve with continuous curvature is called
G^2 continuous.

Suppose the intersection of the two lines $\overline{\mathbf{b}_1, \mathbf{b}_2}$ and $\overline{\mathbf{b}_4, \mathbf{b}_5}$ results
in a point $\mathbf{c}$.[1] Let us define

$$\rho_0 = \text{ratio}(\mathbf{b}_1, \mathbf{b}_2, \mathbf{c}),$$
$$\rho_1 = \text{ratio}(\mathbf{c}, \mathbf{b}_4, \mathbf{b}_5),$$
$$\rho = \text{ratio}(\mathbf{b}_2, \mathbf{b}_3, \mathbf{b}_4).$$

These ratios are easily computed from (1.2). The G^2 condition is now
simply

$$\rho^2 = \rho_0 \rho_1. \tag{9.4}$$

It is illustrated in Sketch 86.

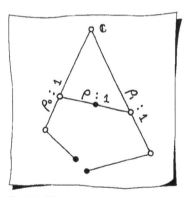

Sketch 86.
Two G^2 cubics.

EXAMPLE 9.3

Using Example 9.1 again, we now change the right curve to

$$\mathbf{b}_3, \mathbf{b}_4, \mathbf{b}_5, \mathbf{b}_6 = \begin{bmatrix} 2 \\ 2 \end{bmatrix}, \begin{bmatrix} 4 \\ 2 \end{bmatrix}, \begin{bmatrix} 4 \\ -2 \end{bmatrix}, \begin{bmatrix} 3 \\ -2 \end{bmatrix}.$$

---

[1]For planar curves, this is trivial (in almost all cases). For space curves, it is
a coplanarity condition for the points $\mathbf{b}_1, \mathbf{b}_2, \mathbf{b}_4, \mathbf{b}_5$.

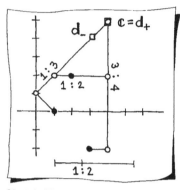

Sketch 87.
A curve that is G^2 but not C^2.

We compute

$$\mathbf{c} = \begin{bmatrix} 4 \\ 5 \end{bmatrix}.$$

Hence

$$\rho_0 = \frac{1}{3}, \quad \rho_1 = \frac{3}{4}, \quad \rho = \frac{1}{2}$$

and since

$$\left(\frac{1}{2}\right)^2 = \frac{1}{3} \times \frac{3}{4},$$

our curve is curvature continuous, or G^2. It is not C^2, however. The points $\mathbf{d}_-$ and $\mathbf{d}_+$ are not equal, as shown in Sketch 87.

Two G^2 curve schemes that have been popular in the computer graphics community are ν-splines and β-splines.

9.4 Working with Piecewise Bézier Curves

Using our G^1 smoothness conditions for Bézier curves, we may now design shapes which are defined by curves. Take the example of the Chinese character "big" which is shown in Figure 9.1. In order to convert this character into piecewise Bézier form, we have to break it down into a number of curve segments, each simple enough to be modeled by a cubic. For this, we pick points on the character's outline which correspond to significant changes in geometry. We mark tangent lines at points where the character is smooth. Where the character has corners, we mark two tangent lines. This is shown in the left part of Figure 9.2.

Order the points as $\mathbf{p}_i$ to reflect the traversal of the character's boundary. With this ordering in mind, the tangent lines must be represented as a vector, indicating this direction. For the method to follow, make these vectors $\mathbf{v}_i$ unit vectors. At junction points where two tangent lines were given, two vectors will be associated with the point; one for the "left" curve and one for the "right" curve. For one cubic segment, set

$$\mathbf{b}_{3i} = \mathbf{p}_i \quad \text{and} \quad \mathbf{b}_{3i+3} = \mathbf{p}_{i+1}.$$

The inner points $\mathbf{b}_{3i+1}$ and $\mathbf{b}_{3i+2}$ will have to be placed on the vectors $\mathbf{v}_i$ and $\mathbf{v}_{i+1}$. Where exactly is a matter of training and trial-and-error,

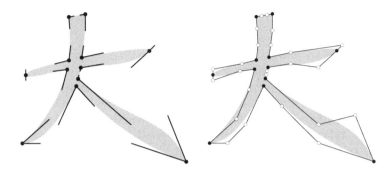

Figure 9.2.
The Chinese character "big." Left, with points and tangents marked. Right, with piecewise Bézier polygon.

but as a rule of thumb one may set

$$\mathbf{b}_{3i+1} = \mathbf{b}_{3i} + 0.4\|\mathbf{b}_{3i+3} - \mathbf{b}_{3i}\|\mathbf{v}_i, \qquad (9.5)$$

$$\mathbf{b}_{3i+2} = \mathbf{b}_{3i+3} - 0.4\|\mathbf{b}_{3i+3} - \mathbf{b}_{3i}\|\mathbf{v}_{i+1}. \qquad (9.6)$$

Figure 9.2, right, shows the resulting piecewise Bézier polygon. It is G^1 at the junctions where one tangent line was specified, and it is G^0 where two tangent lines were specified. We may also look at this problem as one of piecewise Hermite interpolation, see Section 5.6; but the simplicity of the Bézier formulation is far more appealing.

Characters or fonts are often stored as piecewise Bézier curves. This allows for easy rescaling: If a different font size is needed, the Bézier polygons are scaled accordingly. If pixel maps of the fonts were stored, resizing could result in aliasing effects. All letters in this book are generated using PostScript. Thus, each letter in this book is created by evaluating a piecewise Bézier curve!

9.5 Point-Normal Interpolation

In a 3D environment, one is sometimes given a pair of points $\mathbf{p}_0, \mathbf{p}_1 \in \mathbb{E}^3$ and normal vectors $\mathbf{n}_0, \mathbf{n}_1$ at each data point as illustrated in Sketch 88. We would like a cubic which connects $\mathbf{p}_0$ and $\mathbf{p}_1$, with the additional requirement of being tangent to the planes defined by the two normal vectors. Thus we only know the planes in which

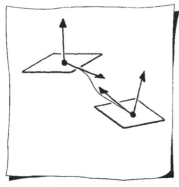

Sketch 88.
Point-normal interpolation.

the curve's tangents have to lie, but we do not know the tangents themselves.

In Bézier form, we already have $b_0 = p_0$ and $b_3 = p_1$. We still need to find b_1 and b_2.

There are infinitely many solutions, so we may try to pick one that is both convenient to compute and of reasonable shape in most cases. As a first approximation to b_1, project b_3 into the plane defined by b_0 and n_0. This defines a tangent at b_0. Place the final b_1 anywhere on this tangent, possibly using the heuristic (9.6). The remaining point b_2 is then obtained analogously.

Applications of this kind of problem are in robotics, where the path of a robot arm is typically described as a piecewise curve. The point-normal pairs are typically extracted from a surface, and the desired curve is intended to lie on the surface—another application of ConS, as discussed in Section 7.5.

9.6 Exercises

1. What are the first and second derivatives at $t = 1$ of the first cubic Bézier curve in Example 9.1? Suppose this curve is now defined over the interval $[0, 2]$. Recompute the first and second derivatives. Sketch the results.

2. Let a linear Bézier curve be given by b_0 and b_1 and a second one by b_1 and b_2. Assume that $b_1 = (1 - \alpha)b_0 + \alpha b_2$ for some $\alpha \in (0, 1)$. What can you say about this composite curve regarding C^1, C^2 and G^1, G^2 continuity?

3. Let a quadratic Bézier curve be given by control points

$$\begin{bmatrix} -2 \\ 0 \end{bmatrix}, \begin{bmatrix} -1 \\ 1 \end{bmatrix}, \begin{bmatrix} 0 \\ 1 \end{bmatrix}$$

and a second one by

$$\begin{bmatrix} 0 \\ 1 \end{bmatrix}, \begin{bmatrix} 1 \\ 1 \end{bmatrix}, \begin{bmatrix} 3 \\ 0 \end{bmatrix}.$$

What can you say about this composite curve regarding C^1, C^2 and G^1, G^2 continuity?

4. Consider two piecewise Bézier curves with control polygon $b_0, \dots, b_6$. Let

$$b_2 = \begin{bmatrix} 0 \\ 1 \end{bmatrix}, b_4 = \begin{bmatrix} 6 \\ 3 \end{bmatrix}.$$

Find the junction point $\mathbf{b}_3$ such that the composite curve is C^1 over the knot sequence $[1, 4, 7]$.

5. Outline the algorithm for finding the intersection of two 3D lines. Be sure to consider numerical issues. Your algorithm may then be used as a building block for the G^2 and C^2 tests of Section 9.3.

6. Consider two piecewise Bézier curves with control polygon $\mathbf{b}_0, \ldots, \mathbf{b}_6$. Let

$$\mathbf{b}_2 = \begin{bmatrix} 1 \\ 0 \end{bmatrix}, \mathbf{b}_3 = \begin{bmatrix} 0 \\ 1 \end{bmatrix}, \mathbf{b}_4 = \begin{bmatrix} -2 \\ 3 \end{bmatrix}.$$

Find a knot sequence such that the piecewise curve is C^1.

7. Let two points be given by

$$\mathbf{p}_0 = \begin{bmatrix} 0 \\ 0 \\ 1 \end{bmatrix}, \quad \mathbf{p}_1 = \begin{bmatrix} 1 \\ 1 \\ 0 \end{bmatrix}.$$

At each point, define a corresponding normal vector

$$\mathbf{n}_0 = \begin{bmatrix} 0 \\ 0 \\ 1 \end{bmatrix}, \quad \mathbf{n}_1 = \begin{bmatrix} 0 \\ 1 \\ 0 \end{bmatrix}.$$

What are the Bézier points for the cubic which solves the point-normal interpolation problem from Section 9.5? Sketch your answer.

8. Sketch—as a piecewise cubic Bézier curve–the outline of the (solid) letter **S**.

9. What is the projection of the point $\begin{bmatrix} 2 \\ 2 \\ 1 \end{bmatrix}$ into the plane defined by the origin and normal $\begin{bmatrix} 0 \\ 1 \\ 0 \end{bmatrix}$?

B-Spline Curves

Figure 10.1.
A B-spline spiral.

Compared with composite Bézier curves, a more complete theory of splines is found in B-spline curves. These days, B-spline curves are often referred to as NURBS (NonUniform Rational B-Splines), which are treated in some length in Chapter 13.

10.1 Basic Definitions

A Bézier curve is defined by

$$\mathbf{x}(t) = \mathbf{b}_0 B_0^n(t) + \ldots + \mathbf{b}_n B_n^n(t).$$

The properties of a Bézier curve are determined by its basis functions B_i^n. Each Bernstein basis function is a polynomial function. B-spline curves are defined by more flexible, *piecewise polynomial* basis functions called B-splines, and give rise to a more general curve method. A B-spline curve is expressed as

$$\mathbf{x}(u) = \mathbf{d}_0 N_0^n(u) + \ldots + \mathbf{d}_{D-1} N_{D-1}^n(u). \qquad (10.1)$$

The N_i^n are the degree n *B-splines*. The precise definition of these piecewise polynomial functions is given in Section 10.5. The $\mathbf{d}_i$ are called *de Boor points* or simply control points.

Let's start with a practical introduction. Figure 10.2 illustrates a variety of cubic B-spline curves. The squares are de Boor points, and the ends of each polynomial curve segment are marked with solid circles. As you can see, the B-spline polygon, formed by the de Boor points, allows for a construction of several polynomial segments. The continuity between pieces can be varied. All of the B-spline polygons are identical; the discussion below will clarify how we get differently shaped curves.

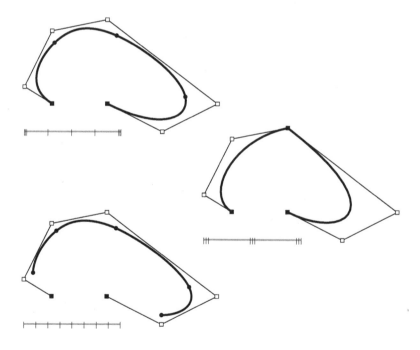

Figure 10.2.
Three cubic B-spline curves.

A degree n B-spline curve is defined by a control polygon

$$\mathbf{d}_0, \ldots, \mathbf{d}_{D-1},$$

similar to a Bézier curve. Because a B-spline curve is a piecewise polynomial curve, we need more than one interval for the parameter— we need a *knot sequence*

$$u_0, \ldots, u_{K-1},$$

which is a sequence of nondecreasing reals. Up to n consecutive knots may coincide, but not more.

The number D of control points is related to the number of K of knots and the degree n by

$$D = K - n + 1.$$

The number of control points D is equal the number of consecutive n-tuples of knots in the knot sequence[1].

When evaluating—plotting—a B-spline curve, only parameter values within the range of knots

$$u_{n-1}, \ldots, u_{K-n}$$

are valid. These are called the *domain knots*. Notice that u_{n-1} is the last knot in the first n-tuple and u_{K-n} is the first knot in the last n-tuple.

As stated above, up to n knots may coincide. The number of coincident values is the *multiplicity*. All of the coincident knots share the multiplicity value. It is common practice to make the first and last n knots of multiplicity n. This results in a curve which passes through the first and last de Boor points, as the top curve in Figure 10.2 does. If a knot has multiplicity one, it is called a *simple knot*. There will be more on this concept in Section 10.3.

If $u_i = u_{i+1}$, then the interval $[u_i, u_{i+1}]$ has length zero. The number of polynomial segments L of a B-spline curve is equal to the number of nonzero length intervals within the domain knots. If all interior domain knots, $u_n, \ldots, u_{K-n-1}$, are simple then $L = K - 2n + 1$. Since D is dependent upon K, we can also, more simply, state that $L = D - n$. See Figure 10.2 for an illustration.

An interval $[u_i, u_{i+m}]$ for $m > 0$ is called a *span* of length m. There are as many spans of length n as there are legs of the control polygon.

---

[1]We'll see why in Section 10.5.

Let's explore the knot sequence and de Boor point relationship with an example.

EXAMPLE 10.1

The top cubic curve in Figure 10.2 has the knot sequence

u_0	u_1	u_2	u_3	u_4	u_5	u_6	u_7	u_8
0	0	0	1	2	3	4	4	4,

thus $K = 9$. This knot sequence is depicted under the curve. The first three knots are of multiplicity three, as is the case for the last three knots. All other knots are simple knots. There are $D = 9 - 3 + 1 = 7$ de Boor points. The domain knots are $u_2, \ldots, u_6$. Each domain knot corresponds to a solid circle on the curve. Recall that the first and last circle correspond to the first and last de Boor point. There are $L = 4$ polynomial segments.

The middle cubic curve in Figure 10.2 has the knot sequence

u_0	u_1	u_2	u_3	u_4	u_5	u_6	u_7	u_8
0	0	0	1	1	1	2	2	2

thus $K = 9$. The first three knots are of multiplicity three, as is the case for the middle and last three knots. Again, $D = 7$ and domain knots are $u_2, \ldots, u_6$. Recall that multiplicity of the knots equal to the degree at the ends causes the curve to pass through the de Boor points. This is also the case in the interior, as illustrated in the figure. This influences the smoothness of the curve segments; more on that in Section 10.6. There are $L = 2$ polynomial segments.

The bottom cubic curve in Figure 10.2 has the knot sequence

u_0	u_1	u_2	u_3	u_4	u_5	u_6	u_7	u_8
0	1	2	3	4	5	6	7	8

thus $K = 9$. All knots are simple. Again, $D = 7$ and the domain knots are $u_2, \ldots, u_6$. Without the multiplicity of the knots equal to the degree at the ends, the curve does not pass through the end de Boor points. There are $L = 4$ polynomial segments.

Some texts add one extra knot at either end of the knot sequence. This is not necessary, but was made popular by a flaw in the data exchange standard IGES.

10.2   The de Boor Algorithm

Bézier curves are evaluated using the de Casteljau algorithm; B-spline curves are evaluated using the *de Boor algorithm*, named after Carl de Boor, who did pioneering work on B-splines. The algorithm uses repeated linear interpolation to compute a point $\mathbf{x}(u)$ on the curve and it proceeds as follows.

Let the evaluation parameter u be within the range of domain knots. Determine the index I such that

$$u_I \leq u < u_{I+1}.$$

In other words, let

$$u \in [u_I, u_{I+1}) \subset [u_{n-1}, u_{K-n}].$$

An exception must be made for $u = u_{K-n}$: set $I = K - n - 1$, the last domain interval.

The de Boor algorithm computes

$$\mathbf{d}_i^k(u) = \frac{u_{i+n-k} - u}{u_{i+n-k} - u_{i-1}} \mathbf{d}_{i-1}^{k-1}(u) + \frac{u - u_{i-1}}{u_{i+n-k} - u_{i-1}} \mathbf{d}_i^{k-1}(u) \quad (10.2)$$

$$\text{for} \quad k = 1, \ldots, n, \quad \text{and}$$
$$i = I - n + k + 1, \ldots, I + 1.$$

The point on the curve is

$$\mathbf{x}(u) = \mathbf{d}_{I+1}^n(u). \tag{10.3}$$

Just as for the de Casteljau algorithm, a convenient schematic tool for describing the algorithm is to arrange the involved points in a triangular diagram:

$$
\begin{array}{cccc}
\mathbf{d}_{I-n+1} & & & \\
\vdots & \mathbf{d}_{I-n+2}^1 & & \\
\vdots & \vdots & & \\
\mathbf{d}_{I+1} & \mathbf{d}_{I+1}^1 & \vdots & \mathbf{d}_{I+1}^n.
\end{array}
\tag{10.4}
$$

The column to the far left consists of the control points of the curve, or $\mathbf{d}_i^0$. Notice that one evaluation involves $n + 1$ de Boor points. This is why B-splines are known for *local control*. The points in successive columns, stages $k = 1, \ldots, n$ will be referred to as intermediate de Boor points. We know they are points since (10.2) is a barycentric combination.

A geometric interpretation of the de Boor algorithm is as follows: (10.2) is simply linear interpolation—and we know this may be viewed as an affine map of the span $[u_{i+n-k}, u_{i-1}]$ to the straight line segment $\mathbf{d}_{i-1}^{k-1}, \mathbf{d}_i^{k-1}$. The point $\mathbf{d}_i^k$ is then the image of u under this affine map.

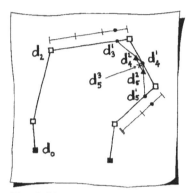

Sketch 89.
The de Boor algorithm.

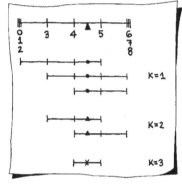

Sketch 90.
De Boor algorithm 's access to the knot sequence.

EXAMPLE 10.2

Sketch 89 illustrates the control polygon $\mathbf{d}_0, \ldots \mathbf{d}_6$ for a cubic B-spline with four segments. The knot sequence for this curve is illustrated in the top of Sketch 90; there are nine knots.

Let's step through the de Boor algorithm for a u parameter that sits in the middle of the third domain interval, marked with a tick mark in Sketch 90. This makes $I = 4$.

Since this is a cubic B-spline, the first ($k = 1$) level of the de Boor algorithm will access spans of three knot intervals. In Sketch 90, all spans of length three which have overlap with $[u_4, u_5]$ have been drawn below the original knot sequence. There are three of them, and together they constitute the first level of the algorithm. Within each span, a circle has been marked to indicate where the parameter u sits.

Each span is mapped to a leg on the polygon. Let's look the first span, which covers $[u_2, u_3, u_4, u_5]$. Looking at the algorithm (10.2), we will find that this span should be mapped to leg $\mathbf{d}_2, \mathbf{d}_3$. Another way to determine which leg is to recall that the ith span of the knot sequence is associated with the ith leg of the control polygon.

This same process,

1. identifying the spans in the knot sequence,

2. marking the relative position of u, and

3. mapping the span to the polygon,

is repeated until the last span covers one interval, $[u_I, u_{I+1}]$. These steps are also depicted in the sketches with different symbols for each stage of the algorithm.

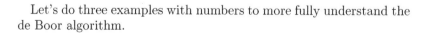

Let's do three examples with numbers to more fully understand the de Boor algorithm.

EXAMPLE 10.3

Let a linear ($n = 1$) B-spline curve be given by the control polygon

$$\begin{bmatrix} -1 \\ 0 \end{bmatrix}, \begin{bmatrix} 0 \\ 1 \end{bmatrix}, \begin{bmatrix} 1 \\ 1 \end{bmatrix}, \begin{bmatrix} 1 \\ 2 \end{bmatrix}$$

and the knot sequence

$$\begin{array}{cccc} 0 & 1 & 2 & 3 \\ u_0 & u_1 & u_2 & u_3. \end{array}$$

We observe that $L = 3$. Let's evaluate this curve at parameter value $u = 1.5$. This parameter value is in the knot interval $[u_1, u_2]$, hence $I = 1$ in the de Boor algorithm. Since the curve is linear, there is only one stage with $i = 2$. We get:

$$\mathbf{d}_2^1(u) = \frac{u_2 - u}{u_2 - u_1}\mathbf{d}_1^0(u) + \frac{u - u_1}{u_2 - u_1}\mathbf{d}_2^0(u)$$

or

$$\mathbf{x}(1.5) = \mathbf{d}_2^1(1.5) = 0.5\begin{bmatrix} 0 \\ 1 \end{bmatrix} + 0.5\begin{bmatrix} 1 \\ 1 \end{bmatrix} = \begin{bmatrix} 0.5 \\ 1 \end{bmatrix}$$

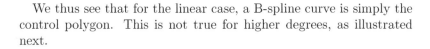

We thus see that for the linear case, a B-spline curve is simply the control polygon. This is not true for higher degrees, as illustrated next.

EXAMPLE 10.4

Let a quadratic ($n = 2$) B-spline curve be given by the control polygon

$$\begin{bmatrix} -1 \\ 0 \end{bmatrix}, \begin{bmatrix} 0 \\ 1 \end{bmatrix}, \begin{bmatrix} 1 \\ 1 \end{bmatrix}, \begin{bmatrix} 1 \\ 2 \end{bmatrix}$$

and the knot sequence

$$\begin{array}{ccccc} 0 & 0 & 1 & 2 & 2 \\ u_0 & u_1 & u_2 & u_3 & u_4. \end{array}$$

We observe that $L = 2$. Let's again evaluate this curve at parameter value $u = 1.5$. The de Boor algorithm is shown in Sketch 91; it shows the individual affine maps involved in the construction. The whole curve is shown in Figure 10.3.

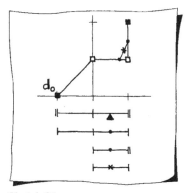

Sketch 91.
The de Boor algorithm for a quadratic curve.

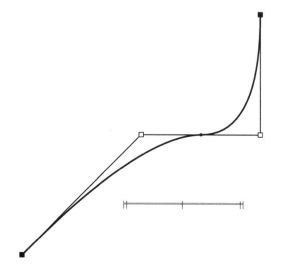

Figure 10.3.
A quadratic B-spline curve with two segments.

Our parameter value is in the knot interval $[u_2, u_3]$, hence $I = 2$ in the de Boor algorithm. We then get

$$\mathbf{d}_2^1(u) = \frac{u_3 - u}{u_3 - u_1}\mathbf{d}_1^0(u) + \frac{u - u_1}{u_3 - u_1}\mathbf{d}_2^0(u),$$

$$\mathbf{d}_3^1(u) = \frac{u_4 - u}{u_4 - u_2}\mathbf{d}_2^0(u) + \frac{u - u_2}{u_4 - u_2}\mathbf{d}_3^0(u),$$

or

$$\mathbf{d}_2^1(1.5) = 0.25 \begin{bmatrix} 0 \\ 1 \end{bmatrix} + 0.75 \begin{bmatrix} 1 \\ 1 \end{bmatrix} = \begin{bmatrix} 0.75 \\ 1 \end{bmatrix},$$

$$\mathbf{d}_3^1(1.5) = 0.5 \begin{bmatrix} 1 \\ 1 \end{bmatrix} + 0.5 \begin{bmatrix} 1 \\ 2 \end{bmatrix} = \begin{bmatrix} 1 \\ 1.5 \end{bmatrix}.$$

The point $\mathbf{d}_3^2$ on the curve is given by

$$\mathbf{d}_3^2(u) = \frac{u_3 - u}{u_3 - u_2}\mathbf{d}_2^1(u) + \frac{u - u_2}{u_3 - u_2}\mathbf{d}_3^1(u),$$

or

$$\mathbf{d}_3^2(1.5) = 0.5 \begin{bmatrix} 0.75 \\ 1 \end{bmatrix} + 0.5 \begin{bmatrix} 1 \\ 1.5 \end{bmatrix} = \begin{bmatrix} 0.875 \\ 1.25 \end{bmatrix}.$$

EXAMPLE 10.5

For a final example, let a cubic $(n = 3)$ B-spline curve be given by
the control polygon

$$\begin{bmatrix} -1 \\ 0 \end{bmatrix}, \quad \begin{bmatrix} 0 \\ 1 \end{bmatrix}, \quad \begin{bmatrix} 1 \\ 1 \end{bmatrix}, \quad \begin{bmatrix} 1 \\ 2 \end{bmatrix}, \quad \begin{bmatrix} 0 \\ 2 \end{bmatrix}$$

and the knot sequence

$$\begin{matrix} 0 & 0 & 0 & 1 & 2 & 2 & 2 \\ u_0 & u_1 & u_2 & u_3 & u_4 & u_5 & u_6. \end{matrix}$$

We observe that $L = 2$. Let's also evaluate this curve at parameter
value $u = 1.5$. The de Boor algorithm is shown in Sketch 92; it shows
the individual affine maps involved in the construction. The whole
curve is shown in Figure 10.4.

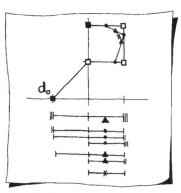

Sketch 92.
The de Boor algorithm for a cubic
curve.

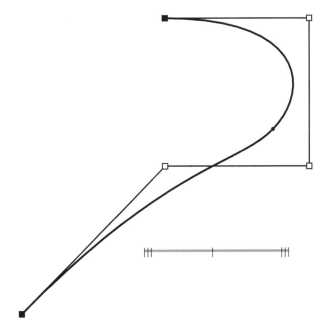

Figure 10.4.
A cubic B-spline curve with two segments.

Our parameter value is in the knot interval $[u_3, u_4]$, hence $I = 3$ in the de Boor algorithm. We get from the first stage of the algorithm:

$$\mathbf{d}_2^1(u) = \frac{u_4 - u}{u_4 - u_1}\mathbf{d}_1^0(u) + \frac{u - u_1}{u_4 - u_1}\mathbf{d}_2^0(u),$$

$$\mathbf{d}_3^1(u) = \frac{u_5 - u}{u_5 - u_2}\mathbf{d}_2^0(u) + \frac{u - u_2}{u_5 - u_2}\mathbf{d}_3^0(u),$$

$$\mathbf{d}_4^1(u) = \frac{u_6 - u}{u_6 - u_3}\mathbf{d}_3^0(u) + \frac{u - u_3}{u_6 - u_3}\mathbf{d}_4^0(u),$$

or

$$\mathbf{d}_2^1(1.5) = 0.25\begin{bmatrix} 0 \\ 1 \end{bmatrix} + 0.75\begin{bmatrix} 1 \\ 1 \end{bmatrix} = \begin{bmatrix} 0.75 \\ 1 \end{bmatrix},$$

$$\mathbf{d}_3^1(1.5) = 0.25\begin{bmatrix} 1 \\ 1 \end{bmatrix} + 0.75\begin{bmatrix} 1 \\ 2 \end{bmatrix} = \begin{bmatrix} 1 \\ 1.75 \end{bmatrix},$$

$$\mathbf{d}_4^1(1.5) = 0.5\begin{bmatrix} 1 \\ 2 \end{bmatrix} + 0.5\begin{bmatrix} 0 \\ 2 \end{bmatrix} = \begin{bmatrix} 0.5 \\ 2 \end{bmatrix}.$$

The points after the second stage are given by

$$\mathbf{d}_3^2(u) = \frac{u_4 - u}{u_4 - u_2}\mathbf{d}_2^1(u) + \frac{u - u_2}{u_4 - u_2}\mathbf{d}_3^1(u),$$

$$\mathbf{d}_4^2(u) = \frac{u_5 - u}{u_5 - u_3}\mathbf{d}_3^1(u) + \frac{u - u_3}{u_5 - u_3}\mathbf{d}_4^1(u),$$

or

$$\mathbf{d}_3^2(1.5) = 0.25\begin{bmatrix} 0.75 \\ 1 \end{bmatrix} + 0.75\begin{bmatrix} 1 \\ 1.75 \end{bmatrix} = \begin{bmatrix} 0.9375 \\ 1.5625 \end{bmatrix},$$

$$\mathbf{d}_4^2(1.5) = 0.5\begin{bmatrix} 1 \\ 1.75 \end{bmatrix} + 0.5\begin{bmatrix} 0.5 \\ 2 \end{bmatrix} = \begin{bmatrix} 0.75 \\ 1.875 \end{bmatrix}.$$

The final point on the curve is given by

$$\mathbf{d}_4^3(u) = \frac{u_4 - u}{u_4 - u_3}\mathbf{d}_3^2(u) + \frac{u - u_3}{u_4 - u_3}\mathbf{d}_4^2(u),$$

or

$$\mathbf{d}_4^3(1.5) = 0.5\begin{bmatrix} 0.9375 \\ 1.5625 \end{bmatrix} + 0.5\begin{bmatrix} 0.75 \\ 1.875 \end{bmatrix} = \begin{bmatrix} 0.84375 \\ 1.71875 \end{bmatrix}.$$

10.3   Practicalities of the de Boor Algorithm

Before making changes to the algorithm, let's take another look at knot multiplicity and a data structure issue. When evaluating a curve for display, you will have to choose an increment which you use to step along the curve. For piecewise polynomials, it is a good idea to specify this increment for each segment, otherwise you might miss a possible corner of your curve. As we have seen, the segments correspond to the non-zero length knot intervals. Since we want to avoid plotting zero-length segments, it is a good idea to label such segments as part of your data structure.

So far, we have described a knot sequence as a one-dimensional floating point array with every knot stored explicitly. This is sometimes called the *expanded knot sequence*. An alternative approach would be to store only the unique floating point values, and then create an integer array which indicates each knot's multiplicity.[2] This array is called the *knot multiplicity vector*. It is also possible to combine the two approaches above, as illustrated in the following example.

EXAMPLE 10.6

We list the knot sequence (value and index) and below it, the multiplicity vector.

0.0	0.0	0.0	1.0	2.0	3.0	3.0	4.0	5.0	5.0	5.0
u_0	u_1	u_2	u_3	u_4	u_5	u_6	u_7	u_8	u_9	u_{10}
3	0	0	1	1	2	0	1	3	0	0

The first knot in a multiplicity is given the multiplicity value, and the other knots are given multiplicity zero.

Another example:

5.0	6.0	10.0	11.0	12.5
u_0	u_1	u_2	u_3	u_4
1	1	1	1	1

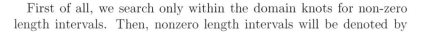

First of all, we search only within the domain knots for non-zero length intervals. Then, nonzero length intervals will be denoted by

---

[2]Be aware that there is a tolerance issue here: When are two knots equal? In order to decide this in a practical situation, a *parameter space tolerance* is needed.

a multiplicity vector value of zero or one, and the next multiplicity vector value must not be equal to zero.

When presented with a parameter value u, we can easily determine which interval it is in and its multiplicity r by accessing the multiplicity vector. Let's simplify the de Boor algorithm to take advantage of this knowledge.

Let

$$u \in [u_I, u_{I+1}) \subset [u_{n-1}, u_{K-n}],$$

and $u_I \neq u_{I+1}$. Again, an exception must be made for $u = u_{K-n}$: set $I = K - n - 1$, the last domain interval. If $u = u_I$ then let r be the multiplicity of u_I, otherwise $r = 0$. The de Boor algorithm computes

$$\mathbf{d}_i^k(u) = \frac{u_{i+n-k} - u}{u_{i+n-k} - u_{i-1}} \mathbf{d}_{i-1}^{k-1}(u) + \frac{u - u_{i-1}}{u_{i+n-k} - u_{i-1}} \mathbf{d}_i^{k-1}(u) \quad (10.5)$$

$$\text{for} \quad k = 1, \ldots, n - r, \quad \text{and}$$
$$i = I - n + k + 1, \ldots, I + 1.$$

The point on the curve is

$$\mathbf{x}(u) = \mathbf{d}_{I+1-r}^{n-r}(u). \quad (10.6)$$

10.4   Properties of B-Spline Curves

We now list several important properties of B-spline curves.

1. *Affine invariance*: If we want to map the B-spline curve by an affine map, all we have to do is to map the control polygon; the map of the B-spline curve is defined by the control polygon image and the original knot sequence.

2. *Differentiability*: At a knot u_i, the curve is $n - 1$ times differentiable—provided that the knot is simple. In the case that the knot is of multiplicity r, the curve is at least $n - r$ times differentiable, or C^{n-r}. The middle curve in Figure 10.2 illustrates a cubic with an internal knot with multiplicity three, and the curve is C^0 there.

3. *Endpoint interpolation*: If $u_0 = \ldots = u_{n-1}$, then $\mathbf{x}(u_{n-1}) = \mathbf{d}_0$, and if $u_{K-n} = \ldots = u_{K-1}$, then $\mathbf{x}(u_{K-n}) = \mathbf{d}_{D-1}$. If the end knots do not have multiplicity n, the curve will not pass through the end control points; see the bottom curve in Figure 10.2 or the left curve of Figure 10.12.

4. *Local control*: If we change a control point $\mathbf{d}_i$, only the closest $n+1$ curve segments change; the curve is left unchanged everywhere else. This is illustrated in Figure 10.5 for three sample curves. Recall that we first identified this property in (10.4).

Local control is what makes B-spline curves more flexible than Bézier curves. If a designer is happy with one part of the curve, subsequent changes in other areas should not affect this part anymore.

5. *Bézier curves*: For some very special knot sequence configurations, B-spline curves are actually Bézier curves. These are knot sequences with $K = 2n$ and $u_0 = \ldots = u_{n-1}$ and $u_n = \ldots = u_{2n-1}$. For the cubic case, such a knot sequence is given by $0, 0, 0, 1, 1, 1$. For these knot sequences, the de Boor algorithm "collapses" to the de Casteljau algorithm. Thus, B-spline curves are a true superset of Bézier curves. See Section 10.6 for more on this relationship.

6. *Endpoint derivatives*: If the knot sequence has end knots of multiplicity n, then the derivatives at the ends take on a very simple form. At the start of the curve, $\dot{\mathbf{x}}(u_{n-1})$ is given by

$$\dot{\mathbf{x}}(u_{n-1}) = \frac{n}{u_n - u_{n-1}}[\mathbf{d}_1 - \mathbf{d}_0]. \tag{10.7}$$

At the other endpoint, we have

$$\dot{\mathbf{x}}(u_{K-n}) = \frac{n}{u_{K-n} - u_{K-n-1}}[\mathbf{d}_{D-1} - \mathbf{d}_{D-2}]. \tag{10.8}$$

Sketch 93 illustrates this for a cubic curve.

7. *Convex hull*: Each point on the curve lies within the convex hull of the control polygon. A stronger convex hull property can also be stated: Each point on the curve lies within the convex hull of no more than $n + 1$ nearby control points.

10.5   B-Splines: The Building Block

B-splines, the basis functions for B-spline curves as expressed in (10.1), are a generalization of Bernstein polynomials. They are composed of several polynomial pieces, instead of being just one polynomial. These pieces fit together such that the B-spline is of a certain smoothness. Figure 10.6 shows two piecewise polynomials: The top

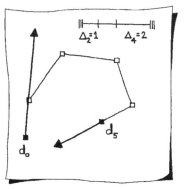

Sketch 93.
Endpoint derivatives for a cubic.

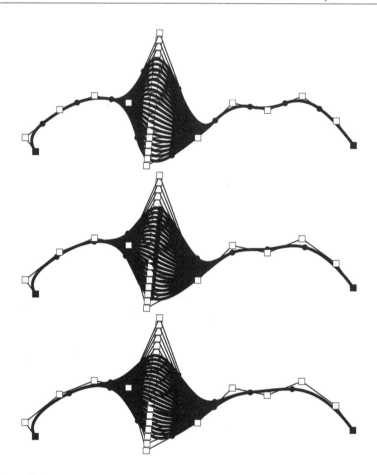

Figure 10.5.

Change of one control point results in local changes of the curve. Curve degrees (from top): $n = 2, 3, 4$. Notice how the affected curve areas become larger as the degree increases.

one is piecewise linear and C^0; the lower one is piecewise quadratic and C^1. Recall from Section 9.2, that the continuity class refers to the differentiability at the junction between polynomials; the polynomials themselves are C^∞.

Figure 10.6 shows the Bézier points of each polynomial segment; the endpoints of each polynomial are marked by solid squares.[3] A

---

[3]For more on the relationship between the B-spline and Bézier forms, see Section 10.6.

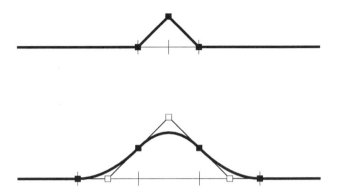

Figure 10.6.
A piecewise linear and a piecewise quadratic function.

B-spline is zero almost everywhere; it assumes nonzero values only for a finite interval. Note how each function vanishes outside a small region; the region where it is nonnegative is called the function's *support*. Our piecewise defined functions were "assembled" as Bézier curves. For higher degrees, this becomes cumbersome, and a more elegant method is called for. This leads to the theory of B-splines, explained next.

We saw in Section 4.8 how to define a functional Bézier curve, i.e., one of the form $y = f(t)$. Degree n *B-spline functions* are defined in a similar way. Their control polygons are given by

$$\mathbf{d}_i = \left[\begin{array}{c} \xi_i \\ d_i \end{array} \right]$$

where $\xi_i = \frac{1}{n}(u_i + \ldots + u_{i+n-1})$. These *Greville abscissae* are the moving averages of the knots—there are as many of them as there are n-tuples of consecutive knots, hence there are as many Greville abscissae as there are control points. The d_i are called the control ordinates of the function.

EXAMPLE 10.7

A cubic B-spline function is shown in Figure 10.7. In that figure, the knot sequence is

$$\begin{array}{cccccccc} 0 & 0 & 0 & 3 & 6 & 12 & 12 & 12 \\ u_0 & u_1 & u_2 & u_3 & u_4 & u_5 & u_6 & u_7. \end{array}$$

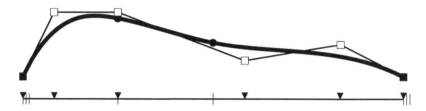

Figure 10.7.
A cubic B-spline function. The control points are positioned over the Greville abscissae.

Thus, the Greville abscissae are given by

$$
\begin{array}{cccccc}
0 & 1 & 3 & 7 & 10 & 12 \\
\xi_0 & \xi_1 & \xi_2 & \xi_3 & \xi_4 & \xi_5
\end{array}
$$

depicted by solid triangular marks.

Notice that the de Boor points are positioned over the Greville abscissae and the curve junction points are positioned over the knots.

B-spline functions are not used too often, with one exception: Suppose for some k, we have $d_k = 1$ and $d_i = 0$ for all other control ordinates. The corresponding B-spline function is called $N_k^n(u)$. Every piecewise polynomial function $f(u)$ may be written as a combination of these B-splines:

$$
f(u) = d_0 N_0^n(u) + \ldots + d_{D-1} N_{D-1}^n(u).
$$

Similarly, every parametric B-spline curve may be written as

$$
\mathbf{x}(u) = \mathbf{d}_0 N_0^n(u) + \ldots + \mathbf{d}_{D-1} N_{D-1}^n(u).
$$

The N_i^n, also called basis splines (or B-splines for short), satisfy the following recursion:

$$
N_i^n(u) = \frac{u - u_{i-1}}{u_{i+n-1} - u_{i-1}} N_i^{n-1}(u) + \frac{u_{i+n} - u}{u_{i+n} - u_i} N_{i+1}^{n-1}(u). \qquad (10.9)
$$

This recursion is anchored by the definition

$$N_i^0(u) = \begin{cases} 1 & \text{if } u_{i-1} \le u < u_i, \\ 0 & \text{else.} \end{cases} \qquad (10.10)$$

In (10.9), if u_j appears that is not part of the knot sequence, e.g., u_{-1}, then assume this term in front of the B-spline function is zero.

In coding this up, one has to be very careful in observing the distinction between the $\le$ and $<$ symbols in (10.10)! Figure 10.8 shows three cubic basis functions defined over the knot sequence in the bottom of the figure.

The recursive definition (10.9) describes each degree n basis function as a linear blend of two degree $n-1$ basis functions. It all starts

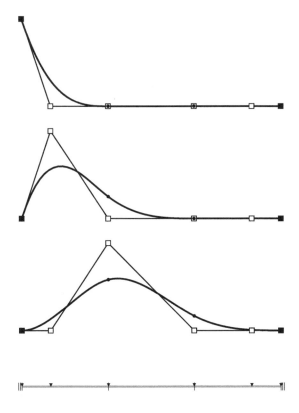

Figure 10.8.
The cubic B-splines N_0^3, N_1^3, and N_2^3 over the given knot sequence.

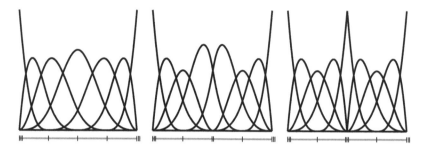

Figure 10.9.
All cubic B-splines over the three given knot sequences. Notice the multiplicity.

with the piecewise constant basis function (10.10); linear basis functions are built from constant, quadratic from linear, etc. Recall that we encountered a similar concept with the de Casteljau algorithm.

We list several properties of B-splines:

1. *Partion of unity*: From the definition of B-splines, it follows that they sum to unity:

$$N_0^n(u) + \ldots + N_{D-1}^n(u) \equiv 1.$$

2. *Linear precision*: If the d_i are sampled at the ξ_i from a linear function $y = au + b$, i.e., $d_i = a\xi_i + b$, then the corresponding B-spline function is that linear function.

3. *Local support*: Every B-spline is nonzero only over $n + 1$ intervals: $N_i^n(u) > 0$ only if $u \in [u_{i-1}, u_{i+n})$.

Figure 10.9 illustrates all cubic basis functions for three knot sequences. The knot sequences differ in the multiplicity of the middle knot.

10.6   Knot Insertion

Knot insertion is a tool for adding a knot to a knot sequence, thereby creating a refined control polygon. The trace of the curve with the refined knot sequence is the same as the original curve. An example, which is illustrated by Sketch 94, will best demonstrate.

EXAMPLE 10.8

Let's begin with a function defined by the knot sequence:

$$
\begin{array}{cccccccc}
0 & 0 & 0 & 1 & 2 & 3 & 3 & 3 \\
u_0 & u_1 & u_2 & u_3 & u_4 & u_5 & u_6 & u_7
\end{array}
$$

Greville abscissae:

$$
\begin{array}{cccccc}
0 & \frac{1}{3} & 1 & 2 & 2\frac{2}{3} & 3 \\
\xi_0 & \xi_1 & \xi_2 & \xi_3 & \xi_4 & \xi_5
\end{array}
$$

and control ordinates $d_0, \ldots, d_5$. The $\mathbf{d}_i = \begin{bmatrix} \xi_i \\ d_i \end{bmatrix}$ are white or black squares in the sketch.

Let's add $u = 1.5$ to the knot sequence, thus we have the new knot sequence:

$$
\begin{array}{ccccccccc}
0 & 0 & 0 & 1 & 1.5 & 2 & 3 & 3 & 3 \\
u_0 & u_1 & u_2 & u_3 & u_4 & u_5 & u_6 & u_7 & u_8
\end{array}
$$

and Greville abscissae:

$$
\begin{array}{ccccccc}
0 & \frac{1}{3} & \frac{5}{6} & \frac{3}{2} & 2\frac{1}{6} & 2\frac{2}{3} & 3 \\
\xi_0 & \xi_1 & \xi_2 & \xi_3 & \xi_4 & \xi_5 & \xi_6
\end{array}
$$

The refined control polygon for this new knot sequence is

$$
\mathbf{d}_0, \quad \mathbf{d}_1, \quad \mathbf{d}_2^1, \quad \mathbf{d}_3^1, \quad \mathbf{d}_4^1, \quad \mathbf{d}_4, \quad \mathbf{d}_5,
$$

and it is illustrated by a solid-line polygon in Sketch 94. The three new points are circles in the sketch. This process of refinement is also known as corner cutting.

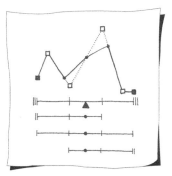

Sketch 94.
Inserting a knot into a functional B-spline curve.

The de Boor algorithm (10.2) is an example of repeated use of knot insertion. Look at the first stage of the algorithm. A parameter u is inserted into the polygon, resulting in a refined polygon. When the knot is inserted n times, we have a point on the curve. However, the de Boor algorithm does not modify the knot sequence or the polygon, leaving it in its original form for the next evaluation. That is why the lengths of the spans decrease at each stage.

A special application of knot insertion is converting from B-spline to piecewise Bézier form. Since B-spline curves are piecewise polynomials, there must be a Bézier polygon for each piece. As we encountered in Section 10.5, the B-spline polygon is actually a Bézier polygon if the knots are of multiplicity n. Hence, all we have to do is to insert each domain knot until it has full multiplicity in order to construct the Bézier points.

EXAMPLE 10.9

Suppose you begin with a cubic B-spline curve with the following knot sequence:

$$\begin{array}{cccccccc} 0 & 0 & 0 & 1 & 2 & 3 & 3 & 3 \\ u_0 & u_1 & u_2 & u_3 & u_4 & u_5 & u_6 & u_7 \end{array}$$

In order to find the refined polygon which is the piecewise Bézier representation of this curve, it is necessary to insert knots so that the knot sequence becomes

$$\begin{array}{cccccccccccc} 0 & 0 & 0 & 1 & 1 & 1 & 2 & 2 & 2 & 3 & 3 & 3 \\ u_0 & u_1 & u_2 & u_3 & u_4 & u_5 & u_6 & u_7 & u_8 & u_9 & u_{10} & u_{11} \end{array}$$

The order in which the knots are inserted doesn't matter. Figure 10.10 illustrates the piecewise Bézier polygons for the three curves

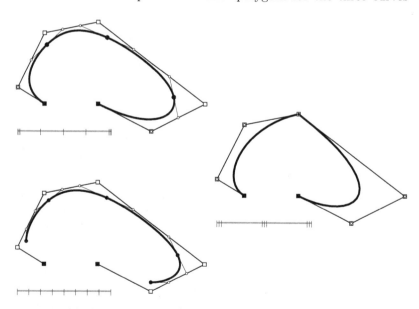

Figure 10.10.

The piecewise Bézier representation of three cubic B-spline curves.

from Figure 10.2. The knot sequence displayed in this figure is for the original curve. The Bézier points are circles. The Bézier polygon approximates the curve more closely, and many calculations are easier for Bézier curves than B-splines.

10.7 Periodic B-Spline Curves

Consider the B-spline curves in Figure 10.11. They show two B-spline curves which are seemingly without beginning or end. Such curves are called *periodic*.

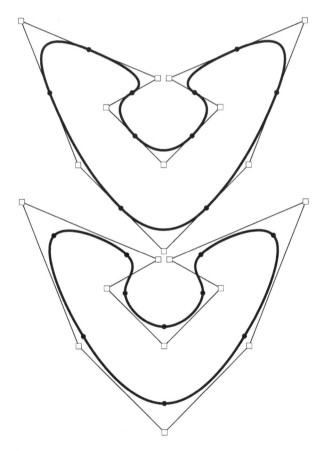

Figure 10.11.
Periodic B-spline curves: top, quadratic, bottom: cubic

A periodic B-spline curve can be constructed as a simple special case of a "normal" one. The goal is to have a seamless control polygon, and evaluation at first domain knot and last domain knot produce the same point. Recall that the de Boor algorithm applied to a particular parameter involves only $n+1$ control points. This gives us the number of control points that must overlap. Taking a closer look at the de Boor algorithm, we can observe that the first $2n - 2$ knot intervals influence the position of the "first" point and the last $2n - 2$ intervals influence the position of the "last" point. Thus the conditions are as follows.

Let $\Delta_i = u_{i+1} - u_i$. Then the knot sequence should be constructed such that we have

$$\Delta_0, \Delta_1, \ldots, \Delta_{2n-3}, \Delta_{2n-2}, \ldots \Delta_{K-2n}, \Delta_0, \Delta_1, \ldots \Delta_{2n-3},$$

and the de Boor points such that

$$\mathbf{d}_0 = \mathbf{d}_{D-n}, \quad \mathbf{d}_1 = \mathbf{d}_{D-(n-1)}, \quad \ldots, \quad \mathbf{d}_{n-1} = \mathbf{d}_{D-1}$$

We illustrate the principle using a simple cubic example. The curve in the left of Figure 10.12 is not "quite" periodic. It uses a knot sequence without multiple end knots. In this case, the knot sequence is $0, 1, 2, 3, 4, 5, 6, 7, 8$. Keep in mind that the curve is only evaluated over the four intervals between $[2, 6]$. The first control point is the solid square in the lower left corner. By forcing the last four control points to be equal to the first four, it becomes periodic as shown in the right of the figure.

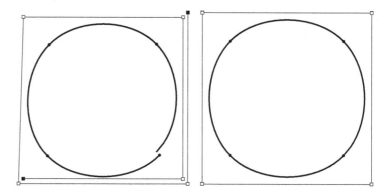

Figure 10.12.

Periodic cubic B-spline curves: left, "almost" periodic, right: truly periodic.

There are other constructions for periodic B-spline curves, however the appealing element in the one above is that your library of B-spline routines will not need a special case for periodic curves. Additionally, if we want to do any geometric analysis of the curve—one time around—the construction lends itself to this just as a "normal" B-spline would.

10.8 Derivatives

By differentiating the N_i^n in (10.1) and manipulating the indices, we arrive at the first derivative for a B-spline curve:

$$\dot{\mathbf{x}}(u) = n[\mathbf{f}_0 N_1^{n-1} + \ldots + \mathbf{f}_{i-1} N_i^{n-1} + \ldots + \mathbf{f}_{D-2} N_{D-1}^{n-1}] \quad (10.11)$$

where

$$\mathbf{f}_{i-1} = \frac{\Delta \mathbf{d}_{i-1}}{u_{n+i-1} - u_{i-1}}.$$

The de Boor algorithm provides an easy way to implement this. The two points $\mathbf{d}_I^{n-1}(u)$ and $\mathbf{d}_{I+1}^{n-1}(u)$ span the curve's tangent:

$$\dot{\mathbf{x}}(u) = \frac{n}{u_{I+1} - u_I}[\mathbf{d}_{I+1}^{n-1}(u) - \mathbf{d}_I^{n-1}(u)]. \quad (10.12)$$

This expression is very similar to the first derivative of a Bézier curve computed via the de Casteljau algorithm. Notice that derivative vectors must be scaled by the length of the parameter interval. In the discussion of piecewise curves in Section 9.1 this was also the case. If the B-spline curve has multiplicity n at the ends, then the derivatives at the ends take a simple form, see (10.7) and (10.8).

The second derivative:

$$\ddot{\mathbf{x}}(u) = n(n-1)[\mathbf{s}_1 N_2^{n-2} + \ldots + \mathbf{s}_{i-1} N_i^{n-2} + \ldots + \mathbf{s}_{D-2} N_{D-1}^{n-2}] \quad (10.13)$$

where

$$\mathbf{s}_{i-1} = \frac{\Delta \mathbf{f}_{i-1}}{u_{n+i-2} - u_{i-1}}.$$

The first derivative formula involved knot sequence spans of length n, and the second derivative involves spans of length $n - 1$.

The second derivative can also be implemented via the de Boor algorithm. Simply compute the intermediate de Boor points up to

$\mathbf{d}_i^{n-2}$ with the normal de Boor algorithm (10.2), and the remaining two steps of the algorithm are modified as follows:

$$\mathbf{d}_i^k(u) = \frac{-k}{u_{i+n-k} - u_{i-1}}\mathbf{d}_{i-1}^{k-1}(u) + \frac{k}{u_{i+n-k} - u_{i-1}}\mathbf{d}_i^{k-1}(u) \quad (10.14)$$

for $\quad k = n - 1, n, \quad$ and
$$i = I - n + k + 1, \ldots, I + 1.$$

Then the second derivative is

$$\ddot{\mathbf{x}}(u) = \mathbf{d}_{I+1}^n(u). \tag{10.15}$$

B-spline formulas tend to look quite complicated. To check if a formula is roughly correct, one tip is to try it out for a B-spline that is simply a Bézier curve. Let's do just that with an example.

EXAMPLE 10.10

Let's start with the cubic curve in Sketch 95 with knot sequence $0, 0, 0, 2, 2, 2$. This is simply one cubic Bézier curve.

We want to test the derivative formulas, but let's evaluate the curve first so we can complete the sketch. Importantly, this will allow us to "guess" the result before we begin computing.

Evaluating this curve at $u = 1.0$, the de Boor algorithm produces the point on the curve:

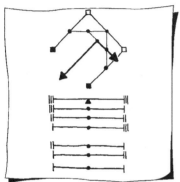

Sketch 95.
A cubic B-spline.

$$
\begin{array}{cccc}
\mathbf{d}_i^0 & \mathbf{d}_i^1 & \mathbf{d}_i^2 & \mathbf{d}_i^3
\end{array}
$$

$$
\begin{bmatrix} -1 \\ 0 \end{bmatrix}
$$
$$
\begin{bmatrix} 0 \\ 1 \end{bmatrix} \quad \begin{bmatrix} -1/2 \\ 1/2 \end{bmatrix}
$$
$$
\begin{bmatrix} 1 \\ 0 \end{bmatrix} \quad \begin{bmatrix} 1/2 \\ 1/2 \end{bmatrix} \quad \begin{bmatrix} 0 \\ 1/2 \end{bmatrix}
$$
$$
\begin{bmatrix} 0 \\ -1 \end{bmatrix} \quad \begin{bmatrix} 1/2 \\ -1/2 \end{bmatrix} \quad \begin{bmatrix} 1/2 \\ 0 \end{bmatrix} \quad \begin{bmatrix} 1/4 \\ 1/4 \end{bmatrix}
$$

The spans for each step of the algorithm are illustrated in the sketch.
The first derivative is computed by differencing the $\mathbf{d}_i^2$:

$$\dot{\mathbf{x}}(1.0) = \frac{3}{2}\left[\begin{bmatrix} 1/2 \\ 0 \end{bmatrix} - \begin{bmatrix} 0 \\ 1/2 \end{bmatrix}\right] = \begin{bmatrix} 3/4 \\ -3/4 \end{bmatrix}.$$

The second derivative begins with the $\mathbf{d}_i^1$ and executes the modified de Boor algorithm (10.14):

$$\begin{bmatrix} -1/2 \\ 1/2 \end{bmatrix}$$

$$\begin{bmatrix} 1/2 \\ 1/2 \end{bmatrix} \quad \begin{bmatrix} 1 \\ 0 \end{bmatrix}$$

$$\begin{bmatrix} 1/2 \\ -1/2 \end{bmatrix} \quad \begin{bmatrix} 0 \\ -1 \end{bmatrix} \quad \begin{bmatrix} -3/2 \\ -3/2 \end{bmatrix},$$

thus

$$\ddot{\mathbf{x}}(1.0) = \begin{bmatrix} -3/2 \\ -3/2 \end{bmatrix}.$$

Another debugging tip: Compute the second derivative for a quadratic curve at a number of parameter values. It should return the same vector for all parameters.

10.9 Exercises

1. Evaluate the curve in Example 10.2 at the midpoint of $[u_3, u_4]$ by sketching the polygon, intermediate control points, knot sequence, and the spans involved in each step of the de Boor algorithm. There is no need to compute numbers.

2. What is the multiplicity vector for the following knot sequence

$$\begin{array}{cccccccc} 0 & 0 & 1 & 3 & 3 & 4 & 5 & 5 \\ u_0 & u_1 & u_2 & u_3 & u_4 & u_5 & u_6 & u_7 \end{array}$$

   If this knot sequence belongs to a quadratic curve, how many segments are there? How many are there if the knot sequence belongs to a cubic curve?

3. What is the derivative at $u = 1.5$ of the B-spline curve from Example 10.5?

4. What is the derivative of the same curve at $u = 0$?

5. Consider the knot sequence $0, 0, 3, 4, 6, 6$. Sketch the B-spline N_2^2. (Hint: First, find the Greville abscissae.)

6. Over the knot sequence $0, 0, 0, 3, 4, 6, 6, 6$ sketch N_1^3.

7. Using the knot sequence from Exercise 5, write the function $y = 3u$ as a quadratic B-spline function.

8. Write the function $y = 3u$ as a cubic B-spline function with knot sequence $0, 0, 0, 3, 4, 6, 6, 6$.

9. The first and second derivative formulas for a B-spline curve in (10.11) and (10.13) follow a particular pattern. In the same manner, write down the third derivative. How would this be implemented via the de Boor algorithm?

10. List the condition(s) for the control polygon of a degree n B-spline curve to be that of a piecewise Bézier curve?

Working with B-Spline Curves

11

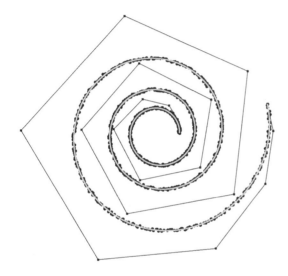

Figure 11.1.
A B-spline approximation to a spiral set of points. Figure courtesy of M. Jeffries.

So far, we have described the basic properties of B-spline curves, but we have not not addressed an important question: How does one use them? B-spline curves owe their popularity to the many possible ways in which they can be "put to work." This chapter explores some of those uses.

11.1 Designing with B-Spline Curves

Imagine you want to create a B-spline curve for the character "v" in some fancy font. A first start could be, using closed cubic B-spline

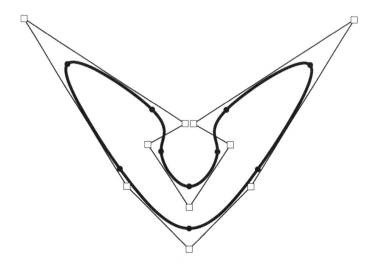

Figure 11.2.

Designing a character with a B-spline curve.

curves, to use the control polygon of the bottom of Figure 10.11. Using a mouse or a similar interactive graphic device, you would move individual control points around until the desired shape is accomplished. You might then arrive at a curve which is shown in Figure 11.2.

This manual method is useful for a final fine tuning of a curve shape; for the complete design process, it is too slow. Some of the methods below provide an initial "guess" which, if necessary, may then be modified interactively.

11.2   Least Squares Approximation

For many applications, many data points are given and a cubic B-spline curve is desired which approximates their shape. The most popular method to find such an approximating curve is that of *least squares approximation*, much in the spirit of a similar method of Section 5.4.

Suppose we want to approximate a data set using a cubic B-spline curve with L polynomial segments. Recall from Section 10.1 that a cubic with only simple domain knots has the relationship

$K = L + 5$, where K is the number of knots. Thus, the first step is to construct a knot sequence $u_0, \ldots, u_{K-1}$. We are given P data points $\mathbf{p}_0, \ldots, \mathbf{p}_{P-1}$, each $\mathbf{p}_i$ being associated with a parameter value v_i. We wish to find a cubic B-spline curve $\mathbf{x}(u)$ such that the distances $\|\mathbf{p}_i - \mathbf{x}(v_i)\|$ are small. Figure 11.3 illustrates the geometry.

Ideally, we would have $\mathbf{p}_i = \mathbf{x}(v_i)$; $i = 0, \ldots, P-1$. If our B-spline curve $\mathbf{x}(u)$ is of the form

$$\mathbf{x}(u) = \mathbf{d}_0 N_0^3(u) + \ldots + \mathbf{d}_{D-1} N_{D-1}^3(u),$$

we would like the following to hold:

$$\mathbf{d}_0 N_0^3(v_0) + \ldots + \mathbf{d}_n N_{D-1}^3(v_0) = \mathbf{p}_0$$

$$\vdots$$

$$\mathbf{d}_0 N_0^3(v_{P-1}) + \ldots + \mathbf{d}_n N_{D-1}^3(v_{P-1}) = \mathbf{p}_{P-1}.$$

This may be condensed into matrix form:

$$\begin{bmatrix} N_0^3(v_0) & \ldots & N_{D-1}^3(v_0) \\ & \vdots & \\ & \vdots & \\ N_0^3(v_{P-1}) & \ldots & N_{D-1}^3(v_{P-1}) \end{bmatrix} \begin{bmatrix} \mathbf{d}_0 \\ \vdots \\ \mathbf{d}_{D-1} \end{bmatrix} = \begin{bmatrix} \mathbf{p}_0 \\ \vdots \\ \vdots \\ \mathbf{p}_{P-1} \end{bmatrix}. \quad (11.1)$$

Or, even shorter:

$$M\mathbf{D} = \mathbf{P}. \quad (11.2)$$

Since we assume the number P of data points is larger than the number D of curve control points, this linear system is clearly *overdetermined*. We attack it, just as in the Bézier case, by simply multiplying both sides by M^{T}:

$$M^{\mathrm{T}}M\mathbf{D} = M^{\mathrm{T}}\mathbf{P}. \quad (11.3)$$

This is a linear system with D equations in D unknowns, with a square and symmetric coefficient matrix $M^{\mathrm{T}}M$. Its solution is straightforward since $M^{\mathrm{T}}M$ is always invertible.[1]

An example is shown in Figure 11.3. One thousand data points were placed on a spiral and some noise was added.

Our development lacked some details that are essential when implementing "real life" examples. How many segments L should the

---

[1]This is true only if the parameter values v_j are distributed without large gaps throughout the knot sequence $\{u_j\}$.

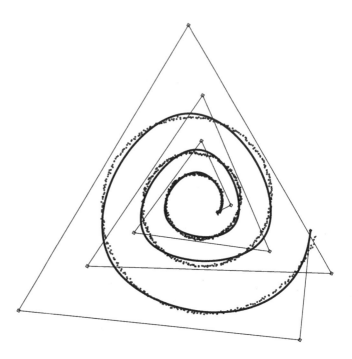

Figure 11.3.
Least squares approximation to a spiral using a cubic B-spline curve. Figure
courtesy of M. Jeffries.

curve have? How should the knots u_j and the parameter values v_i be
chosen?

There are no universal answers to these problems. But you might
want to consider the following:

- Choose the parameters v_i according to the chord length method
  explained in Section 5.5.

- Select $L \approx P/10$.

- Choose the knots u_i such that approximately ten v_j fall in each
  interval domain knot interval $[u_i, u_{i+1}]$.

11.3   Shape Equations

If the data points are very noisy, very unevenly distributed, or otherwise "nasty," then straightforward least squares approximation might fail to produce nice results. One cure is the following.

In the least squares formulation, there is no account for the shape of the resulting curve. We might be willing to deviate a bit from the data if that would buy a better-shaped curve. Reasoning about the curve's shape is hard—it involves concepts such as curvature, as described in Section 8.2. It is easier to formulate conditions for the control polygon's shape. The shape of a polygon is nice if it does not "wiggle" much. This may be expressed by computing *differences* of the control points.

A common measure for polygon shape is the use of second differences. These are of the form

$$\Delta^2 \mathbf{d}_i = \mathbf{d}_i - 2\mathbf{d}_{i+1} + \mathbf{d}_{i+2}.$$

A polygon is considered "nice" if the sum

$$\|\Delta^2 \mathbf{d}_0\| + \ldots + \|\Delta^2 \mathbf{d}_{D-3}\|$$

is small.

The following example should give credibility to this concept.

EXAMPLE 11.1

Let a polygon be given by the 2D points

$$\begin{bmatrix} 0 \\ 0 \end{bmatrix}, \begin{bmatrix} 1 \\ 1 \end{bmatrix}, \begin{bmatrix} 2 \\ 0 \end{bmatrix}, \begin{bmatrix} 3 \\ 1 \end{bmatrix}, \begin{bmatrix} 4 \\ 0 \end{bmatrix}$$

and a second one by

$$\begin{bmatrix} 0 \\ 0 \end{bmatrix}, \begin{bmatrix} 1 \\ 1 \end{bmatrix}, \begin{bmatrix} 2 \\ 1.5 \end{bmatrix}, \begin{bmatrix} 3 \\ 1 \end{bmatrix}, \begin{bmatrix} 4 \\ 0 \end{bmatrix}.$$

Both polygons are shown in Sketch 96; visually, the first one is "rougher" than the second one.

We compute the second differences of the first polygon:

$$\Delta^2 \mathbf{d}_0 = \begin{bmatrix} 0 \\ -2 \end{bmatrix}, \quad \Delta^2 \mathbf{d}_1 = \begin{bmatrix} 0 \\ 2 \end{bmatrix}, \quad \Delta^2 \mathbf{d}_2 = \begin{bmatrix} 0 \\ -2 \end{bmatrix}.$$

For the second polygon, we find

$$\Delta^2 \mathbf{d}_0 = \begin{bmatrix} 0 \\ -0.5 \end{bmatrix}, \quad \Delta^2 \mathbf{d}_1 = \begin{bmatrix} 0 \\ -1 \end{bmatrix}, \quad \Delta^2 \mathbf{d}_2 = \begin{bmatrix} 0 \\ -0.5 \end{bmatrix}.$$

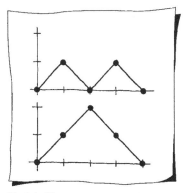

Sketch 96.
A rough polygon, top and a smooth one, bottom.

The sum of the lengths of the first polygon's difference vectors is six; that of the second polygon is only two: the second one is smoother! This confirms that second differences tell us something about the shape of a polygon.

Returning to the topic of least squares approximation, it seems reasonable to add *shape equations* to the overdetermined system (11.2). These would be of the form

$$\mathbf{d}_0 - 2\mathbf{d}_1 + \mathbf{d}_2 = \mathbf{0}$$

$$\vdots$$

$$\mathbf{d}_{D-3} - 2\mathbf{d}_{D-2} + \mathbf{d}_{D-1} = \mathbf{0}.$$

With the addition of these equations, our overdetermined linear system becomes even more overdetermined. However, this is not detrimental at all—we still form the normal equations of the form (11.3) and solve them for the control points. Figure 11.4 illustrates the effect of the shape equations.

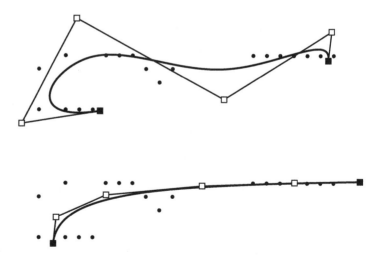

Figure 11.4.
The effect of shape equations. Top: without shape equations, and below: with shape equations.

11.4   Cubic Spline Interpolation

The most popular way of constructing cubic spline curves is through interpolation. This means that the number of data items (also called *interpolation constraints*) equals that of the unknown control points.

The task at hand is to interpolate to P given data points $\mathbf{p}_0, \ldots, \mathbf{p}_{P-1}$ with a cubic B-spline curve which has end knots of multiplicity three:

$$u_0 = u_1 = u_2, \quad u_3, \ldots, u_{K-4}, \quad u_{K-3} = u_{K-2} = u_{K-1}.$$

The junction points of the curve will be paired to the given data points; for example $\mathbf{p}_0$ will correspond to u_2, $\mathbf{p}_1$ will correspond to u_3, etc. Therefore, we will need $P - 1$ curve segments, and thus $K = P + 4$. A method for determining the knots based on the given data can be found in Section 5.5.

Because of the relationship between the number of knots and control points, the interpolating curve has $D = P + 2$ control points, and needs $P + 2$ data items to determine it. Example 11.2 should give you a feeling for the number of control points and curve segments.

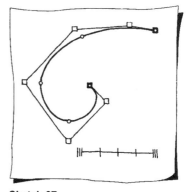

EXAMPLE 11.2

Suppose we have $P = 5$ data points, and we consider the knot sequence

$$0, 0, 0, 1, 2, 3, 4, 4, 4$$

with $K = 5 + 4 = 9$ knots, thus having five junction points[2].

A control polygon for a cubic B-spline curve over this knot sequence is given by the $D = 7$ control points

$$\mathbf{d}_0, \ldots, \mathbf{d}_6.$$

Sketch 97.
A cubic B-spline curve with four cubic segments.

Thus, one needs seven interpolation constraints to determine the curve. Sketch 97 shows a configuration like the one just described.

The example above shows that two more data items are needed than the curve has junction points (or will have, since we don't know

---

[2]This includes the beginning and end as junction points.

it yet). Typically, one "throws in" two more data items at the ends of the curve, namely the derivatives

$$\mathbf{t}_s = \dot{\mathbf{x}}(u_2) \quad \text{and} \quad \mathbf{t}_e = \dot{\mathbf{x}}(u_{K-3}).$$

Here, $\mathbf{t}_s$ and $\mathbf{t}_e$ stand for the start and end tangents of the curve. These are called *end conditions*. The knots u_2 and u_{K-3} are the first and last domain knots, and simple expressions for these derivatives are given in (10.8) and (10.7). A good method for generating these tangents is called *Bessel tangents*. These tangents are extracted from the interpolating parabolas through the first and last three data points. They are given by

$$\mathbf{t}_s = -\frac{2\Delta_2 + \Delta_3}{\Delta_2(\Delta_2 + \Delta_3)}\mathbf{p}_0 + \frac{(\Delta_2 + \Delta_3)}{\Delta_2\Delta_3}\mathbf{p}_1 - \frac{\Delta_2}{\Delta_3(\Delta_2 + \Delta_3)}\mathbf{p}_2 \quad (11.4)$$

$$
\begin{aligned}
\mathbf{t}_e &= \frac{\Delta_{K-4}}{\Delta_{K-5}(\Delta_{K-5} + \Delta_{K-4})}\mathbf{p}_{L-2} \\
&\quad - \frac{(\Delta_{K-5} + \Delta_{K-4})}{\Delta_{K-5}\Delta_{K-4}}\mathbf{p}_{L-1} \\
&\quad + \frac{(2\Delta_{K-4} + \Delta_{K-5})}{(\Delta_{K-5} + \Delta_{K-4})\Delta_{K-4}}\mathbf{p}_L
\end{aligned}
\qquad (11.5)
$$

If $P = 2$ then set

$$\mathbf{t}_s = \mathbf{t}_e = \mathbf{p}_1 - \mathbf{p}_0$$

in order to get the linear interpolant.

Now we have as many knowns as unknowns, namely $D = P + 2$. Our interpolation conditions become

$$
\begin{aligned}
\mathbf{p}_0 &= \mathbf{x}(u_2) \\
\mathbf{t}_s &= \dot{\mathbf{x}}(u_2) \\
\mathbf{p}_1 &= \mathbf{x}(u_3) \\
&\vdots \\
\mathbf{t}_e &= \dot{\mathbf{x}}(u_{K-3}) \\
\mathbf{p}_{P-1} &= \mathbf{x}(u_{K-3}).
\end{aligned}
\qquad (11.6)
$$

However, as you might have noticed in Sketch 97, cubic spline interpolation with triple end knots will result in

$$\mathbf{d}_0 = \mathbf{p}_0 \quad \text{and} \quad \mathbf{d}_{D-1} = \mathbf{p}_{P-1}.$$

This clearly eliminates two unknowns, and therefore, we can eliminate two equations, and (11.6) becomes

$$\mathbf{t}_s = \dot{\mathbf{x}}(u_2)$$
$$\mathbf{p}_1 = \mathbf{x}(u_3)$$
$$\vdots \tag{11.7}$$
$$\mathbf{p}_{P-2} = \mathbf{x}(u_{K-4})$$
$$\mathbf{t}_e = \dot{\mathbf{x}}(u_{K-3}),$$

for the $P = D - 2$ unknowns $\mathbf{d}_1, \ldots, \mathbf{d}_{D-2}$.

EXAMPLE 11.3

Let us reconsider Example 11.2. The interpolation conditions (11.6) are given by

$$\mathbf{p}_0 = \mathbf{x}(0)$$
$$\mathbf{t}_s = \dot{\mathbf{x}}(0)$$
$$\mathbf{p}_1 = \mathbf{x}(1)$$
$$\mathbf{p}_2 = \mathbf{x}(2)$$
$$\mathbf{p}_3 = \mathbf{x}(3)$$
$$\mathbf{t}_e = \dot{\mathbf{x}}(4)$$
$$\mathbf{p}_4 = \mathbf{x}(4).$$

By assigning $\mathbf{d}_0 = \mathbf{p}_0$ and $\mathbf{d}_6 = \mathbf{p}_4$, thus eliminating the first and last equations, the system becomes

$$\mathbf{t}_s = \dot{\mathbf{x}}(0)$$
$$\mathbf{p}_1 = \mathbf{x}(1)$$
$$\mathbf{p}_2 = \mathbf{x}(2)$$
$$\mathbf{p}_3 = \mathbf{x}(3)$$
$$\mathbf{t}_e = \dot{\mathbf{x}}(4),$$

which are five equations for the unknowns $\mathbf{d}_1, \ldots, \mathbf{d}_5$.

In theory, each data point yields an equation of the form

$$\mathbf{p}_i = \mathbf{d}_0 N_0^3(u_{2+i}) + \ldots + \mathbf{d}_{D-1} N_{D-1}^3(u_{2+i}).$$

But due to the local support property of B-spline curves, this reduces to

$$\mathbf{p}_i = \mathbf{d}_i N_i^3(u_{2+i}) + \mathbf{d}_{i+1} N_{i+1}^3(u_{2+i}) + \mathbf{d}_{i+2} N_{i+2}^3(u_{2+i}). \qquad (11.8)$$

This gives the system a *tridiagonal* structure.[3] The first and last equation in the system, the end conditions, involve only the first and last unknowns, respectively, in order to maintain this structure.

For the special case of equally spaced interior knots, we have

$$6\mathbf{p}_i = \mathbf{d}_i + 4\mathbf{d}_{i+1} + \mathbf{d}_{i+2}$$

for each equation involving a data point.

EXAMPLE 11.4

Returning to our example, notice that the knots are equally spaced. The end tangent equations are

$$\mathbf{t}_s = 3(\mathbf{d}_1 - \mathbf{d}_0) \quad \text{and} \quad \mathbf{t}_e = 3(\mathbf{d}_6 - \mathbf{d}_5),$$

where $\mathbf{d}_0$ and $\mathbf{d}_6$ are known. Therefore, the linear system is

$$\begin{bmatrix} 1 & 0 & 0 & 0 & 0 \\ \frac{3}{2} & \frac{7}{2} & 1 & 0 & 0 \\ 0 & 1 & 4 & 1 & 0 \\ 0 & 0 & 1 & \frac{7}{2} & \frac{3}{2} \\ 0 & 0 & 0 & 0 & 1 \end{bmatrix} \begin{bmatrix} \mathbf{d}_1 \\ \mathbf{d}_2 \\ \mathbf{d}_3 \\ \mathbf{d}_4 \\ \mathbf{d}_5 \end{bmatrix} = \begin{bmatrix} \mathbf{d}_0 + \frac{1}{3}\mathbf{t}_s \\ 6\mathbf{p}_1 \\ 6\mathbf{p}_2 \\ 6\mathbf{p}_3 \\ \mathbf{d}_6 - \frac{1}{3}\mathbf{t}_e \end{bmatrix}$$

An example is illustrated in Figure 11.5.

11.5   Cubic Spline Interpolation in a Nutshell

Let's summarize cubic spline interpolation from Section 11.4.
Input: data points $\mathbf{p}_0, \ldots, \mathbf{p}_{P-1}$ and a cubic B-spline knot sequence

$$u_0 = u_1 = u_2, \quad u_3, \ldots, u_{K-4}, \quad u_{K-3} = u_{K-2} = u_{K-1}$$

where $K = P + 4$, thus there are $P - 1$ segments.

---

[3]Thus a special tridiagonal solver should be used which is much faster than employing a general purpose system solver.

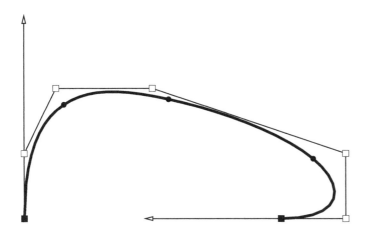

Figure 11.5.

Cubic spline interpolation.

Task: Find the control points $\mathbf{d}_0, \ldots, \mathbf{d}_{D-1}$, where $D = P + 2$, for the cubic B-spline interpolant to the given data with the given knot sequence. Each data point $\mathbf{p}_i$ is associated with parameter u_{2+i}. Compute:

- Set $\mathbf{d}_0 = \mathbf{p}_0$ and $\mathbf{d}_{D-1} = \mathbf{p}_{P-1}$.

- Create tangents $\mathbf{t}_s$ and $\mathbf{t}_e$ using Bessel tangents (11.4) and (11.5), respectively.

- Set up the tridiagonal linear system of equations (11.7), where (11.8) completes the description of the middle equations, and the first and last equations are

$$\mathbf{d}_0 + \frac{\Delta_2}{3}\mathbf{t}_s = \mathbf{d}_1$$

$$\mathbf{d}_{D-1} - \frac{\Delta_{K-4}}{3}\mathbf{t}_e = \mathbf{d}_{D-2}$$

- Solve the $(D-2) \times (D-2)$ linear system for $\mathbf{d}_1, \ldots, \mathbf{d}_{D-2}$.

11.6 Exercises

1. Design an algorithm to choose the knot sequence for least squares approximation if the data points and their parameters are given.

2. What is a chord length knot sequence for the following data points?

$$\begin{bmatrix} 0 \\ 0 \end{bmatrix} \begin{bmatrix} 2 \\ 0 \end{bmatrix} \begin{bmatrix} 2 \\ 2 \end{bmatrix} \begin{bmatrix} 3 \\ 2 \end{bmatrix} \begin{bmatrix} 4 \\ 0 \end{bmatrix}$$

3. What are the Bessel tangents for the following data points, assuming uniform parameters?

$$\begin{bmatrix} 0 \\ 0 \end{bmatrix}, \begin{bmatrix} 3 \\ 1 \end{bmatrix}, \begin{bmatrix} 6 \\ 2 \end{bmatrix}.$$

4. Repeat with these data points:

$$\begin{bmatrix} 0 \\ 0 \end{bmatrix}, \begin{bmatrix} 3 \\ 1 \end{bmatrix}, \begin{bmatrix} 6 \\ 0 \end{bmatrix}.$$

5. Find the cubic spline interpolant with a uniform knot sequence for the previous data set, using Bessel tangents.

6. Use the data from the previous problem, but replace the Bessel tangents by the explicit derivative vectors

$$\mathbf{t}_s = \begin{bmatrix} -1 \\ -1 \end{bmatrix}, \quad \mathbf{t}_e = \begin{bmatrix} 1 \\ 1 \end{bmatrix}.$$

Find the cubic spline interpolant.

Composite Surfaces

12

© Pixar Animation Studios

Figure 12.1.
A scene from Pixar's "Geri's Game", modeled using subdivision surfaces.

In this chapter, we cover surfaces composed of more than one patch. One Bézier patch is rarely flexible enough to model a real life part. More commonly, many patches will be stitched together, either in the form of a composite Bézier surface or a B-spline surface. In animation, subdivision surfaces are a popular type of composite surface. Figure 12.1 is taken from "Geri's Game", an animated short film produced by Pixar Animation Studios.

12.1 Composite Bézier Surfaces

In Chapter 9, we saw how to generate complex curves which consisted of simple ingredients, i.e., cubic Bézier curves. In the context of surfaces, the same approach is taken. Once we know how to combine several Bézier patches to form a smooth composite surface, we can build surfaces of fairly high complexity.

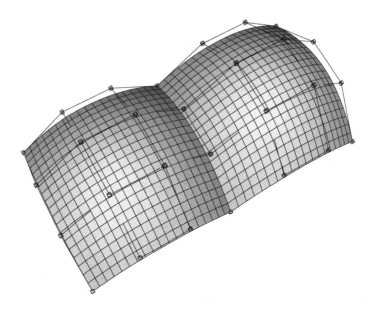

Figure 12.2.
Two adjacent C^0 bicubic patches.

Let $\mathbf{b}_{i,j}; 0 \leq i, j \leq 3$ be a bicubic Bézier patch—we'll call it the "left" patch. Let $\mathbf{b}_{i,j}; 3 \leq i \leq 6; 0 \leq j \leq 3$ be a second ("right") patch such that both share a common boundary. As is typical for piecewise defined objects, we make some assumptions regarding the domains of our patches. We assume the left patch has as its domain a rectangle $[u_0, u_1] \times [v_0, v_1]$. For the right patch, we assume that the domain is a rectangle $[u_1, u_2] \times [v_0, v_1]$. Thus, the two domain rectangles satisfy the same neighboring information as holds for the patches, namely they share a common $u = 1$ boundary. See Figure 12.2.

We are now interested in conditions regarding the smoothness of the composite surface formed by the left and right patch. These conditions are surprisingly simple. We observe that the composite control net contains four rows of control points:

$$\mathbf{b}_{0,0}, \ldots, \mathbf{b}_{6,0}$$
$$\mathbf{b}_{0,1}, \ldots, \mathbf{b}_{6,1}$$
$$\mathbf{b}_{0,2}, \ldots, \mathbf{b}_{6,2}$$
$$\mathbf{b}_{0,3}, \ldots, \mathbf{b}_{6,3}$$

Each of these rows of control points may be interpreted as the piece-wise Bézier polygon of a composite cubic curve over the knot sequence u_0, u_1, u_2. Our surface is C^1 if all these rows satisfy the curve C^1 conditions from Section 9.1. This is very simple to check and to work with!

For our bicubics, these conditions are

$$\mathbf{b}_{3,j} = \frac{\Delta_1}{\Delta}\mathbf{b}_{2,j} + \frac{\Delta_0}{\Delta}\mathbf{b}_{4,j}; \quad j = 0, 1, 2, 3. \qquad (12.1)$$

where $\Delta_0 = u_1 - u_0$, $\Delta_1 = u_2 - u_1$, and $\Delta = u_2 - u_0$.

In words: Any three points $\mathbf{b}_{2,j}, \mathbf{b}_{3,j}, \mathbf{b}_{4,j}$ must be collinear *and* they must be in the same ratio:

$$\mathrm{ratio}(\mathbf{b}_{2,j}, \mathbf{b}_{3,j}, \mathbf{b}_{4,j}) = \frac{\Delta_0}{\Delta_1}.$$

An example of a composite bicubic C^1 surface is shown in Figure 12.3.

If the degrees of our patches are not bicubic, we obtain completely analogous conditions, as we will see in the next example.

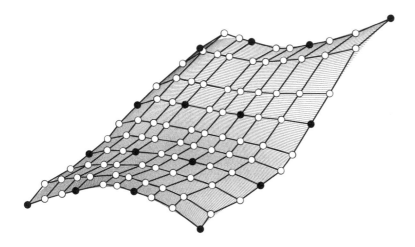

Figure 12.3.
A composite smooth surface. Knot sequences: $u_i : 0, 1, 3, 4$; $v_j : 0, 1, 2, 3$.
(The u parameter corresponds to the "horizontal" curves.)

EXAMPLE 12.1

Let a bilinear patch be given by control points

$$
\left[\begin{array}{cc}
\begin{bmatrix} 0 \\ 0 \\ 0 \end{bmatrix} & \begin{bmatrix} 1 \\ 0 \\ 0 \end{bmatrix} \\
\begin{bmatrix} 0 \\ 1 \\ 0 \end{bmatrix} & \begin{bmatrix} 1 \\ 1 \\ 1 \end{bmatrix}
\end{array} \right],
$$

and it is defined over $[0,1] \times [0,1]$. A second bilinear patch is defined by

$$
\left[\begin{array}{cc}
\begin{bmatrix} 1 \\ 0 \\ 0 \end{bmatrix} & \begin{bmatrix} 4 \\ 0 \\ 0 \end{bmatrix} \\
\begin{bmatrix} 1 \\ 1 \\ 1 \end{bmatrix} & \begin{bmatrix} 2 \\ 1 \\ 2 \end{bmatrix}
\end{array} \right],
$$

and it is defined over $[0,1] \times [1,2]$. They meet along the $v = 1$ boundary.

We observe that the two rows of control points

$$
\begin{bmatrix} 0 \\ 0 \\ 0 \end{bmatrix} \begin{bmatrix} 1 \\ 0 \\ 0 \end{bmatrix} \begin{bmatrix} 4 \\ 0 \\ 0 \end{bmatrix}
$$

and

$$
\begin{bmatrix} 0 \\ 1 \\ 0 \end{bmatrix} \begin{bmatrix} 1 \\ 1 \\ 1 \end{bmatrix} \begin{bmatrix} 2 \\ 1 \\ 2 \end{bmatrix}
$$

do indeed contain three collinear points each. However, the ratios are different: We have a ratio of $1:3$ for the first triple of points and a ratio of $1:1$ for the second triple.

Hence, the composite surface cannot be C^1. A direct way of showing this is by computing the derivatives $\mathbf{x}_u(1, 0.5)$ for both patches. See Exercise 1 for details.

We see that C^1 conditions for composite surfaces are quite simple to handle. However, if we have a rectangular network of patches, there is a price to be paid for this: Since there are only two knot sequences—one for the u-direction and one for the v-direction—there is a certain amount of inflexibility. If not all u-isoparametric curves have similar shapes, then a common knot sequence for all of them is not a good idea. The same holds for the v-curves.

12.2   B-Spline Surfaces

Recall how we generalized Bézier curves to Bézier patches: This cumulated in the matrix form (6.7) for a point on a Bézier patch which we repeat here:

$$\mathbf{x}(u,v) = \begin{bmatrix} B_0^m(u) & \dots & B_m^m(u) \end{bmatrix} \begin{bmatrix} \mathbf{b}_{0,0} & \dots & \mathbf{b}_{0,n} \\ \vdots & & \vdots \\ \mathbf{b}_{m,0} & \dots & \mathbf{b}_{m,n} \end{bmatrix} \begin{bmatrix} B_0^n(v) \\ \vdots \\ B_n^n(v) \end{bmatrix}. \quad (12.2)$$

It was derived from the matrix form of a Bézier curve (4.13) which we also repeat here:

$$\mathbf{x}(t) = N^{\mathrm{T}}\mathbf{B}. \quad (12.3)$$

Recall the definition of a B-spline curve; it was given in (10.1):

$$\mathbf{x}(u) = \mathbf{d}_0 N_0^n(u) + \dots + \mathbf{d}_{D-1} N_{D-1}^n(u). \quad (12.4)$$

In matrix form, we can abbreviate this as

$$\mathbf{x}(u) = N^{\mathrm{T}}\mathbf{D}. \quad (12.5)$$

It should now be fairly straightforward to take this to the definition of a B-spline surface $\mathbf{x}(u,v)$:

$$\mathbf{x}(u,v) = \begin{bmatrix} N_0^m(u) & \dots & N_{D-1}^m(u) \end{bmatrix} \begin{bmatrix} \mathbf{d}_{0,0} & \dots & \mathbf{d}_{0,E-1} \\ \vdots & & \vdots \\ \mathbf{d}_{D-1,0} & \dots & \mathbf{d}_{D-1,E-1} \end{bmatrix} \begin{bmatrix} N_0^n(v) \\ \vdots \\ N_{E-1}^n(v) \end{bmatrix},$$
$$(12.6)$$

which we may abbreviate to a more compact

$$\mathbf{x}(u,v) = M^{\mathrm{T}}\mathbf{D}N. \quad (12.7)$$

This definition requires a knot sequence each in the u- and v-direction. They are given by

$$u_0, u_1, \dots, u_{R-1},$$
$$v_0, v_1, \dots, v_{S-1}.$$

Figure 12.4 shows a bicubic B-spline surface.

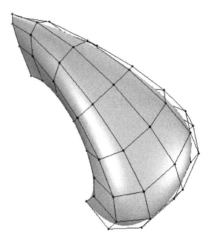

Figure 12.4.

A bicubic B-spline surface. over the knot sequences $u_i = 0, 1, 2, 3, 4, 5$ and $v_j = 0, 1, 2, 3, 4$.

B-spline surfaces enjoy all the properties of Bézier patches, such as symmetry, affine invariance, and convex hull property.

There are two major differences, however. The boundary polygons of the control mesh will be control polygons of the boundary curves only if the end knots have full multiplicity m and n. This is analogous to the endpoint interpolation property of B-spline curves.

Secondly, B-spline surfaces have the *local control* property: If one control point is moved, then only up to $(m + 1)(n + 1)$ patches in its vicinity are affected. Figure 12.5 illustrates. The differences between the two surfaces appear through *Moiré patterns*. The apparent "waves" are due to this effect, and are not part of either surface.

An important operation is the extraction of B-spline control points of an *isoparametric curve*. Let's pick a curve with constant parameter value $u = \bar{u}$. We follow the same approach that we used for (6.9) and (6.10).

Setting

$$\mathbf{C} = M^{\mathrm{T}}\mathbf{D} = [\mathbf{c}_0, \ldots, \mathbf{c}_{E-1}],$$

we factor (12.7) as

$$\mathbf{x}(\bar{u}, v) = \mathbf{C}N.$$

This last equation is that of a B-spline curve with variable v and control polygon $\mathbf{c}_0, \ldots, \mathbf{c}_{E-1}$; as v varies, $\mathbf{x}(\bar{u}, v)$ traces out the desired isocurve.

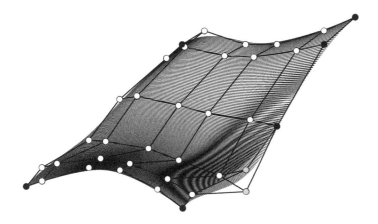

Figure 12.5.
Local control: Two control nets differ by only one control point, marked in gray for either net. The two surfaces only differ in the vicinity of the differing points.

Factoring the other way, i.e., first forming $\mathbf{D}N$, gives rise to isocurves with $v = \bar{v}$ held fixed. Once an isocurve control polygon is found, it may be evaluated and/or differentiated, following the material in Chapter 10.

A B-spline surface consists of a collection of individual polynomial patches, namely $(D - m)(E - n)$ of them if all interior domain knots are simple. Each of these patches may be written in Bézier form, i.e., as patches of degrees m in u and of degree n in v. How do we obtain these individual patch control nets? We first convert each row of control points into piecewise Bézier form. Then we convert each column of the result into piecewise Bézier form again. These conversion steps are described in Section 10.6. Figure 12.6 shows a bicubic B-spline surface together with its piecewise bicubic Bézier control nets.

Figure 12.7 shows a different surface, obtained from the same B-spline control net but with a different knot sequence in the u-direction.

12.3 B-Spline Surface Approximation

B-spline surfaces are more general than Bézier patches, but many applications follow exactly the framework that we already encountered in Chapter 7. The least squares approximation technique is a

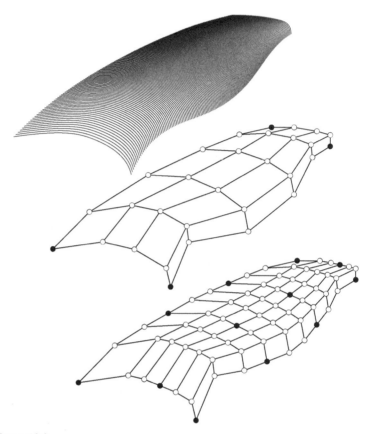

Figure 12.6.
A bicubic B-spline surface. Top, the surface; middle, its B-spline control polygon;
bottom, its piecewise Bézier contol net. Knot sequences: uniform in u and v.

good example. See Section 7.6 for a description of the corresponding
methods for Bézier patches.

We are given data points $\mathbf{p}_k$; $k = 0, \ldots, K - 1$ and we assume that
each data point $\mathbf{p}_k$ has a corresponding pair of parameters (u_k, v_k).
These parameter pairs are expected to be in the domain of a B-spline
surface which we would like to determine. In other words, we assume
that we know the two knot sequences and the u- and v-degrees of
the desired approximating surface. Once we know that, everything is
near trivial.

We wish to find a B-spline surface that fits the data as good as possible. We again introduce the linearized ordering of terms as discussed

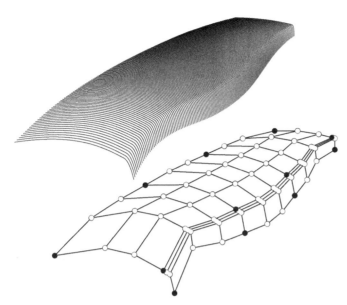

Figure 12.7.
A bicubic B-spline surface. Top, the surface; bottom, its piecewise Bézier contol net.
Knot sequences: $u_i : 0, 1, 3, 4$; $v_j : 0, 1, 2, 3$.

in Section 7.6. We then have

$$\mathbf{x}(u, v) = \left[\; N_0^m(u)N_0^n(v), \ldots, N_{D-1}^m(u)N_{E-1}^n(v) \; \right] \begin{bmatrix} \mathbf{d}_{0,0} \\ \vdots \\ \mathbf{d}_{D-1,E-1} \end{bmatrix}.$$

For the kth data point $\mathbf{p}_k$, this becomes $\mathbf{p}_k = \mathbf{x}(u_k, v_k)$ or

$$\mathbf{x}(u, v) = \left[\; N_0^m(u_k)N_0^n(v_k), \ldots, N_{D-1}^m(u_k)N_{E-1}^n(v_k) \; \right] \begin{bmatrix} \mathbf{d}_{0,0} \\ \vdots \\ \mathbf{d}_{D-1,E-1} \end{bmatrix}.$$

Combining all K of these equations, we obtain

$$\begin{bmatrix} \mathbf{p}_0 \\ \vdots \\ \vdots \\ \vdots \\ \mathbf{p}_{K-1} \end{bmatrix} = \begin{bmatrix} N_0^m(u_0)N_0^n(v_0) & \ldots & N_{D-1}^m(u_0)N_{E-1}^n(v_0) \\ & \vdots & \\ & \vdots & \\ & \vdots & \\ & \vdots & \\ N_0^m(u_{K-1})N_0^n(v_{K-1}) & \ldots & N_{D-1}^m(u_{K-1})N_{E-1}^n(v_{K-1}) \end{bmatrix} \begin{bmatrix} \mathbf{d}_{0,0} \\ \vdots \\ \mathbf{d}_{D-1,E-1} \end{bmatrix}.$$

We abbreviate this to $\qquad \mathbf{P} = M\mathbf{D}.$ $\hfill (12.8)$

The least squares solution is found by solving

$$M^{\mathrm{T}}\mathbf{P} = M^{\mathrm{T}}M\mathbf{D}, \qquad (12.9)$$

which is the familiar system of *normal equations*.

In many situations, the parameter values (u_k, v_k) are not known a priori, and hence have to be estimated. The techniques in Section 7.6 provide help.

The solution to the least squares problem may have unsatisfactory shape in some cases or may not even be solvable in cases where the data exhibit "holes" in some regions. Both problems are cured by the introduction of *shape equations*. The motivation is as follows: If each quadrilateral of the control mesh were a parallelogram, we would have a "nice" mesh. It would satisfy

$$\mathbf{d}_{i,j} + \mathbf{d}_{i+1,j+1} - \mathbf{d}_{i+1,j} - \mathbf{d}_{i,j+1} = \mathbf{0} \qquad (12.10)$$

for each quadrilateral. We now simply add each of the equations formed from (12.10) to the overdetermined system (12.8) and then solve this new, more overdetermined, system using the normal equations. This use of shape equations is the analogy to those employed in Section 11.3.

An example of a B-spline surface least squares fit is given in Figure 12.8. The application at hand is a shoe last.[1]

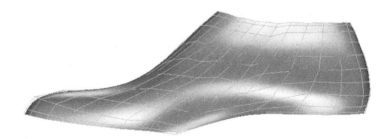

Figure 12.8.
A least squares surface fit to a shoe last.

---

[1] A shoe last is the piece of wood or plastic on which the leather for a shoe is stretched; it is then left to dry so that it assumes its final shape.

12.4 B-Spline Surface Interpolation

In the context of bicubic B-spline surfaces, we may also formulate an interpolation problem as follows. Suppose we are given a $P \times Q$ rectangular array of data points $\mathbf{p}_{i,j}$. We wish to find an interpolating bicubic B-spline surface such that the corners of the patches go through the given data points.

Building on the curve problem in Section 11.4, the surface will have knot sequences u_i for $0 \leq i < R$ and v_i for $0 \leq i < S$. where $R = P+4$ and $S = Q + 4$. The control net

$$\mathbf{d}_{i,j} \qquad 0 \leq i < D, \quad 0 \leq j < E,$$

must have $D = P + 2$ and $E = Q + 2$ control points.

We may solve a surface interpolation problem by reducing it to a series of curve interpolation problems. We interpret the given array of data points as a set of P rows of points. To each of these rows with Q points, we fit a B-spline curve using the algorithm from Section 11.5, resulting in $Q + 2$ control points in each row. This, in turn, results in a $P \times (Q + 2)$ net of control points $\mathbf{c}_{i,j}$. This array of points is now treated in a column-by-column fashion: To each of these $(Q+2)$ columns, we fit a B-spline curve through P points, again using the algorithm from Section 11.5, resulting in $P + 2$ control points in each column. This forms the final result, a $(P + 2) \times (Q + 2)$ control net $\mathbf{d}_{i,j}$ of the desired surface!

EXAMPLE 12.2

Let's start with a 2×3 array of data points

$$\begin{bmatrix} \mathbf{p}_{0,0} & \mathbf{p}_{0,1} & \mathbf{p}_{0,2} \\ \mathbf{p}_{1,0} & \mathbf{p}_{1,1} & \mathbf{p}_{1,2} \end{bmatrix}.$$

There are two rows of data points, each with three points. To each row, we fit a B-spline curve, resulting in two control polygons:

$$\begin{bmatrix} \mathbf{c}_{0,0} & \mathbf{c}_{0,1} & \mathbf{c}_{0,2} & \mathbf{c}_{0,3} & \mathbf{c}_{0,4} \\ \mathbf{c}_{1,0} & \mathbf{c}_{1,1} & \mathbf{c}_{1,2} & \mathbf{c}_{1,3} & \mathbf{c}_{1,4} \end{bmatrix}.$$

Now we treat each of the five columns as a set of curve data points. The Bessel tangents from Section 11.5 will produce a cubic control polygon for each, resulting in

$$\begin{bmatrix} \mathbf{d}_{0,0} & \mathbf{d}_{0,1} & \mathbf{d}_{0,2} & \mathbf{d}_{0,3} & \mathbf{d}_{0,4} \\ \mathbf{d}_{1,0} & \mathbf{d}_{1,1} & \mathbf{d}_{1,2} & \mathbf{d}_{1,3} & \mathbf{d}_{1,4} \\ \mathbf{d}_{2,0} & \mathbf{d}_{2,1} & \mathbf{d}_{2,2} & \mathbf{d}_{2,3} & \mathbf{d}_{2,4} \\ \mathbf{d}_{3,0} & \mathbf{d}_{3,1} & \mathbf{d}_{3,2} & \mathbf{d}_{3,3} & \mathbf{d}_{3,4} \end{bmatrix}.$$

Note that these five polygons correspond to five linear curves (degree elevated to cubic) since we defined Bessel tangents to produce linear segments if only two data points are given. The $\mathbf{d}_{i,j}$ form the desired surface control mesh.

Note that in treating each of the row and column sets of problems, we have to solve tridiagonal linear systems with only one matrix for all row problems and only one matrix for all column problems. This reduces the amount of work considerably!

12.5 Subdivision Surfaces: Doo-Sabin

The de Casteljau algorithm and the de Boor algorithm are two examples of schemes which refine a polygon to the point that the polygon locally approximates a smooth curve. Both algorithms are actually repeated instances of knot insertion (see Section 10.6).

Chaikin's algorithm is an application of this refinement concept. The input is simply a polygon and the output is a refined polygon which approximates a smooth curve. In Figure 12.9, the input polygon is depicted by square points. The bottom part of the figure illustrates the final refined polygon. Suppose the input polygon is $\mathbf{d}_0, \mathbf{d}_1, \ldots, \mathbf{d}_n$. One step of Chaikin's algorithm produces the refined polygon $\mathbf{d}_0^1, \mathbf{d}_1^1, \ldots, \mathbf{d}_{2n-1}^1$, where

$$\mathbf{d}_0^1 = \mathbf{d}_0$$
$$\mathbf{d}_{2i-1}^1 = \frac{3}{4}\mathbf{d}_i + \frac{1}{4}\mathbf{d}_{i-1}$$
$$\mathbf{d}_{2i}^1 = \frac{3}{4}\mathbf{d}_i + \frac{1}{4}\mathbf{d}_{i+1}$$
$$\mathbf{d}_{2n-1}^1 = \mathbf{d}_n$$

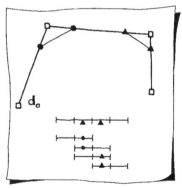

Sketch 98.
One step of Chaikin's algorithm.

for $i = 1, \ldots, n-1$. This algorithm can be repeatedly applied until we obtain a polygon which is smooth, as in the bottom of Figure 12.9. The top of the figure illustrates two steps of the algorithm with each step in a different shade of grey. Notice the "corner cutting" nature of this scheme.

Looking a little closer at Chaikin's algorithm, it becomes apparent that it is a special application of knot insertion. The input polygon consists of de Boor points $\mathbf{d}_i$ of a quadratic B-spline. The knot sequence for this B-spline is uniform. Sketch 98 illustrates that one step

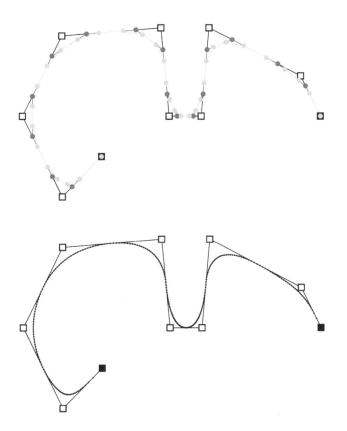

Figure 12.9.
Chaikin's algorithm applied to the polygon with square vertices. Top: two steps.
Bottom: five steps.

of the algorithm is equivalent to inserting a knot at the midpoint of each domain knot interval.

As we have seen with tensor product surfaces, many curve concepts can be carried over by simply starting with a control net with rectangular structure, and assuming there are two knot sequences. Chaikin's algorithm for curves can be generalized in just this manner, resulting in the *Doo-Sabin algorithm*, which converges to biquadratic B-splines defined over uniform knots. However, the Doo-Sabin algorithm is more than that. The algorithm generalizes Chaikin's so that it can be applied to polygonal meshes of arbitrary topology; in other words, the polygons do not have to be be four-sided.

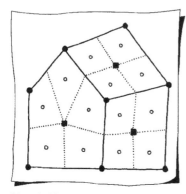

Sketch 99.
Forming new vertices in the
Doo-Sabin algorithm.

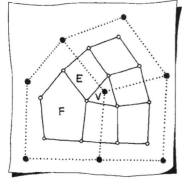

Sketch 100.
Forming new faces in the Doo-
Sabin algorithm.

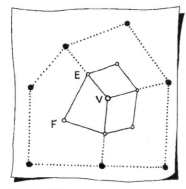

Sketch 101.
The Catmull-Clark algorithm.

One step of the Doo-Sabin algorithm consists of the following.

1. Form new vertices: A face with n vertices will produce n new vertices. Sketch 99 illustrates the elements of this step, which is done for each face.

   a. Form the face's centroid: the average of the vertices.

   b. Form the edge midpoints of the face.

   c. Each new vertex is the average of a face vertex, the face centroid, and two adjacent edge midpoints.

2. Form the new faces from the new vertices. There are three types of constructions, illustrated in Sketch 100.

   a. F-faces are constructed by joining new vertices within a face.

   b. E-faces are constructed by joining new vertices at the edges of neighboring old faces.

   c. V-faces are constructed by joining all new vertices around an old vertex.

Repeat until the polygonal mesh is of the desired smoothness. Figure 12.10 illustrates the algorithm.

A substructure consisting of four quadrilaterals, whose vertices form a 3×3 rectangular net, corresponds to the control net of a biquadratic B-spline patch over uniform knot sequences. Neighboring rectangular patches are C^1. When the faces are not four-sided, the surface will in general be less smooth there.[2] Non-four-sided faces appear if they are in the input mesh, and also if the input mesh has $n \neq 4$ faces around a vertex. In this latter case, the first step of Doo-Sabin creates a face with n vertices. Non-four-sided faces are not created in later stages of the algorithm. As each non-four-sided face shrinks to a point, it is called an *extraordinary vertex*.

12.6   Subdivision Surfaces: Catmull-Clark

Generalizing a cubic curve subdivision scheme to produce bicubic B-splines results in the *Catmull-Clark algorithm*. Just as the Doo-Sabin algorithm, this algorithm has been further generalized to work on polygonal meshes of arbitrary topology.

One step of the Catmull-Clark algorithm consists of the following steps, and is illustrated in Sketch 101.

---

[2]The parametric continuity concept can be applied only for rectangular geometry.

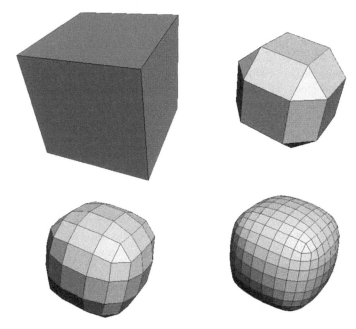

Figure 12.10.
Several levels of the Doo-Sabin algorithm. Courtesy of A. Nasri.

1. Form face points **f**: For each face in the mesh, average the face's vertices.

2. Form edge points **e**: For each edge in the mesh, average the edge's vertices and the two face points on either side of the edge.

3. Form vertex points **v**: For each vertex in the mesh, form a weighted average of three elements; there are n faces around a vertex. A vertex point is defined as

$$\mathbf{v} = [\frac{(n-3)}{n}](\text{old vertex point})$$
$$+ [\frac{1}{n}](\text{average of } \mathbf{f}'s \text{ around } \mathbf{v})$$
$$+ [\frac{2}{n}](\text{average of midpoints of edges around } \mathbf{v})$$

4. Form the faces of the new mesh. Each face consists of four vertices of the type $(\mathbf{f}, \mathbf{e}, \mathbf{v}, \mathbf{e})$, connected in that order. (All faces after the first level of subdivision will be four-sided.)

A substructure consisting of nine quadrilaterals whose vertices form a 4×4 rectangular net, corresponds to the control net of a bicubic B-spline surface over uniform knot sequences. Neighboring rectangular surfaces are C^2. Catmull-Clark surface's extraordinary vertices differ from those of Doo-Sabin. The first step of Catmull-Clark produces all four-sided faces. However, n faces will share a vertex if an input face was n-sided, or if the input mesh had n-faces around a vertex. This non-rectangular element will shrink with more steps, but will not go away. The continuity is diminished at these extraordinary vertices.

Subdivision surfaces are appealing for a number of reasons. First, by starting with a simple polygonal net, a smooth surface of the same general shape can quickly and easily be generated. The topology issue is a big plus. For example, consider something as simple as a sphere. Rectangular patches stitched together will necessitate a degenerate patch to close it up. For these reasons, the graphics and animation industries have embraced subdivision surfaces—see Figure 12.1. Figure 12.11 illustrates the algorithm.

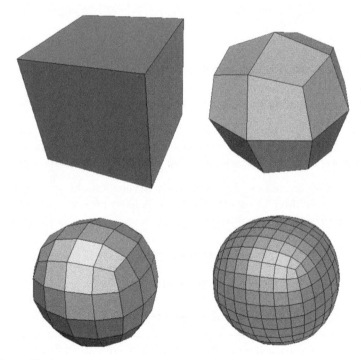

Figure 12.11.
Several steps of the Catmull-Clark algorithm. Courtesy of A. Nasri.

12.7   Exercises

1. Assume that the composite surface of Example 12.1 has $u_0, u_1, u_2 = 0, 1, 2$ and $v_0, v_1 = 0, 1$ as its domain. Compute $\mathbf{x}_u(1, 0.5)$, once from the "left" and once from the "right."

2. To what position do you have to change the point $[4, 0, 0]$ on the "right" surface such that you obtain a C^1 surface?

3. Let four data points $\mathbf{p}_{i,j}$ be given by

$$
\left[
\begin{array}{cc}
\begin{bmatrix} 0 \\ 0 \\ 0 \end{bmatrix} & \begin{bmatrix} 6 \\ 0 \\ 0 \end{bmatrix} \\
\begin{bmatrix} 0 \\ 6 \\ 0 \end{bmatrix} & \begin{bmatrix} 6 \\ 6 \\ 6 \end{bmatrix}
\end{array}
\right].
$$

What is the control net of the interpolating bicubic B-spline surface? (This does not depend on the knot sequences.)

4. Then repeat for data points $\mathbf{p}_{i,j}$ given by

$$
\left[
\begin{array}{cc}
\begin{bmatrix} 0 \\ 0 \\ 0 \end{bmatrix} & \begin{bmatrix} 6 \\ 0 \\ 0 \end{bmatrix} \\
\begin{bmatrix} 0 \\ 6 \\ 0 \end{bmatrix} & \begin{bmatrix} 6 \\ 12 \\ 12 \end{bmatrix}
\end{array}
\right].
$$

5. Let four control points of a bilinear patch be given by

$$
\left[
\begin{array}{cc}
\mathbf{b}_{0,0} & \mathbf{b}_{1,0} \\
\mathbf{b}_{0,1} & \mathbf{b}_{1,1}
\end{array}
\right]
=
\left[
\begin{array}{cc}
\begin{bmatrix} 0 \\ 0 \\ 0 \end{bmatrix} & \begin{bmatrix} 4 \\ 0 \\ 0 \end{bmatrix} \\
\begin{bmatrix} 0 \\ 4 \\ -2 \end{bmatrix} & \begin{bmatrix} 6 \\ 6 \\ 2 \end{bmatrix}
\end{array}
\right].
$$

Let the domain for the corresponding Bézier patch be $[0, 1] \times [0, 1]$. Let a second bilinear Bézier patch be defined over $[0, 1] \times [1, 3]$ with Bézier points

$$
\left[
\begin{array}{cc}
\mathbf{b}_{0,1} & \mathbf{b}_{1,1} \\
\mathbf{b}_{0,2} & \mathbf{b}_{1,2}
\end{array}
\right].
$$

If both patches are to form a C^1 surface, what are the Bézier points $\mathbf{b}_{0,2}$ and $\mathbf{b}_{1,2}$?

6. The Doo-Sabin algorithm is derived by considering knot insertion at the midpoint of each domain interval. Each new vertex formed involves only the vertices of one face. If the input mesh is a rectangular net, what are the *masks* that express each of the four new vertices in terms of the face vertices? Reduce your work by keeping symmetry in mind!

NURBS 13

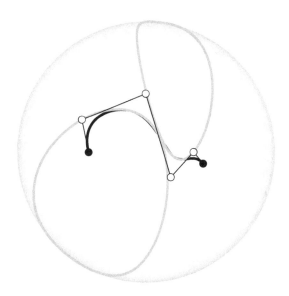

Figure 13.1.
Conics in the circular model of the projective plane.

We shall now describe the main geometric features of this curve and surface representation. Much of the discussion of B-spline curves in Chapter 10 and B-spline surfaces in Chapter 12 applies to NURBS. This chapter focuses on the special features of NURBS. Surprisingly, most of these features are already exhibited by conics, which are a special case of NURBS.

13.1 Conics

Conic sections, the oldest known curve form, are still essential to many CAD systems. In fact, conics were used as the basis for the first "CAD" system: This was done by R. Liming [18] in 1944—he based the design of airplane fuselages upon *calculating* with conics, as opposed to the traditional *drafting* with conics.

A number of equivalent ways exist to define a conic section; for our purposes the following one is very useful: *A conic section in $I\!\!E^2$ is the perspective projection of a parabola in $I\!\!E^3$.*

When it comes to the formulation of conics as rational curves, one typically chooses the center of the projection to be the origin $\mathbf{0}$ of a 3D Cartesian coordinate system. The plane into which one projects is taken to be the plane $z = 1$. Since we will study planar curves in this section, we may think of this plane as a copy of $I\!\!E^2$. Our special projection is characterized by

$$\underline{\mathbf{x}} = \begin{bmatrix} x \\ y \\ z \end{bmatrix} \longrightarrow \begin{bmatrix} x/z \\ y/z \end{bmatrix} = \mathbf{x}.$$

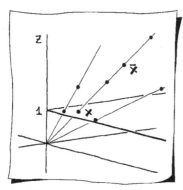

Sketch 102.
Homogeneous coordinates.

Note that a point $\mathbf{x}$ is the projection of a whole family of points: Every point of the form $f\underline{\mathbf{x}}$ projects onto the same 2D point $\mathbf{x}$. The 3D point $\underline{\mathbf{x}}$ is called the *homogeneous form* or *homogeneous coordinates* of $\mathbf{x}$. Sketch 102 illustrates.

A conic $\mathbf{c}(t)$ is given by *weights* $z_0, z_1, z_2 \in I\!\!R$ and *control points* $\mathbf{b}_0, \mathbf{b}_1, \mathbf{b}_2 \in I\!\!E^2$ such that

$$\mathbf{c}(t) = \frac{z_0\mathbf{b}_0 B_0^2(t) + z_1\mathbf{b}_1 B_1^2(t) + z_2\mathbf{b}_2 B_2^2(t)}{z_0 B_0^2(t) + z_1 B_1^2(t) + z_2 B_2^2(t)}, \tag{13.1}$$

i.e., $\mathbf{c}$ may be expressed as a *parametric rational quadratic curve.*

Thus the conic control polygon is the projection of the control polygon with homogenoues vertices

$$z_0 \begin{bmatrix} \mathbf{b}_0 \\ 1 \end{bmatrix}, \quad z_1 \begin{bmatrix} \mathbf{b}_1 \\ 1 \end{bmatrix}, \quad z_2 \begin{bmatrix} \mathbf{b}_2 \\ 1 \end{bmatrix},$$

which is the control polygon of the 3D parabola that we projected onto the conic $\mathbf{c}$. The form (13.1) is called the *rational quadratic form* of a conic section.

EXAMPLE 13.1

Let three homogeneous control points be given by

$$\underline{\mathbf{b}}_0 = \begin{bmatrix} 0 \\ 1 \\ 1 \end{bmatrix}, \quad \underline{\mathbf{b}}_1 = \begin{bmatrix} 2 \\ 2 \\ 2 \end{bmatrix}, \quad \underline{\mathbf{b}}_2 = \begin{bmatrix} 4 \\ 0 \\ 2 \end{bmatrix}.$$

They project onto the 2D points

$$\mathbf{b}_0 = \begin{bmatrix} 0 \\ 1 \end{bmatrix}, \quad \mathbf{b}_1 = \begin{bmatrix} 1 \\ 1 \end{bmatrix}, \quad \mathbf{b}_2 = \begin{bmatrix} 2 \\ 0 \end{bmatrix}$$

for which we record weights $z_0, z_1, z_2 = 1, 2, 2$. At $t = 0.5$, we have

$$\underline{\mathbf{x}}(0.5) = \begin{bmatrix} 2 \\ 5/4 \\ 7/4 \end{bmatrix} \quad \longrightarrow \quad \begin{bmatrix} 8/7 \\ 5/7 \end{bmatrix} = \mathbf{x}(0.5).$$

For an illustration, see Sketch 103.

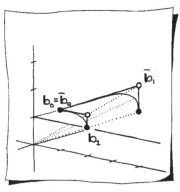

Sketch 103.
A point on a conic.

13.2 Reparametrization and Classification

It is possible to change the weights of a conic without changing its shape. If the initial weights are z_0, z_1, z_2, then the set of weights $z_0, cz_1, c^2 z_2$ generates the same conic for any $c \neq 0$. This may be used to bring a conic into *standard form*: Assuming $z_0 = 1$ without loss of generality, we set $c = 1/\sqrt{z_2}$.[1] Now the new weights are $1, cz_1, 1$.

EXAMPLE 13.2

For the weights $1, 2, 2$ of the conic from Example 13.1, we find $c = 1/\sqrt{2}$. The new weights in standard form are

$$1, \frac{2}{\sqrt{2}}, 1.$$

Changing the weights in this fashion does not change the curve's geometry, but it does change how it is traversed. Hence, the term *reparametrization* is used to describe this process.

Once a conic is in standard form, it is easy to decide which type it is:

- a hyperbola if $z_1 > 1$;

- a parabola if $z_1 = 1$;

- an ellipse if $z_1 < 1$.

---

[1] We can achieve $z_0 = 1$ by dividing all weights by z_0.

Figure 13.2.

As the weight z_1 changes from 0.1 to 10, three types of conics are produced.

Figure 13.2 shows some examples. The "flat" segments are ellipses, while the "curved" ones are hyperbolas. The intermediate case, the parabola, is plotted in gray.

An interesting effect occurs if we set $c = -1$. Then, the weights z_0, z_1, z_2 change to new weights $z_0, -z_1, z_2$. While the first set of weights (assuming all z_i are positive) generates a curve inside the control polygon, the second set generates the remaining "half" of the curve, called the *complementary segment*. This comes in handy if we want to plot a whole conic: Simply plot a conic arc for each set of weights. This is done in Figure 13.3.

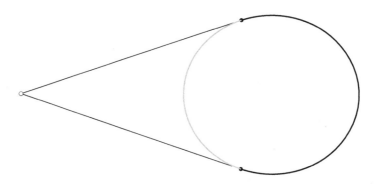

Figure 13.3.

An arc of a conic, grey, and the complementary segment, black.

13.3 Derivatives

We write a conic section as a rational function, so derivatives now call for the quotient rule—at least at first sight. Life is actually easier than that, using a little trick. Our conic $\mathbf{c}(t)$ is of the form $\mathbf{c}(t) = \mathbf{p}(t)/z(t)$, with a polynomial numerator $\mathbf{p}(t)$. Thus

$$\mathbf{p}(t) = z(t)\mathbf{c}(t).$$

This polynomial curve is differentiated using the product rule:

$$\dot{\mathbf{p}}(t) = \dot{z}(t)\mathbf{c}(t) + z(t)\dot{\mathbf{c}}(t),$$

the dot denoting differentiation with respect to t. The expression $\dot{\mathbf{c}}(t)$ on the right hand side is our desired conic derivative! We can solve for it and obtain

$$\dot{\mathbf{c}}(t) = \frac{1}{z(t)}[\dot{\mathbf{p}}(t) - \dot{z}(t)\mathbf{c}(t)]. \tag{13.2}$$

This procedure may be repeated to yield higher order derivatives.

Let us now consider two conics, one defined over the interval $[u_0, u_1]$ with control polygon $\mathbf{b}_0, \mathbf{b}_1, \mathbf{b}_2$ and weights w_0, w_1, w_2 and the other defined over the interval $[u_1, u_2]$ with control polygon $\mathbf{b}_2, \mathbf{b}_3, \mathbf{b}_4$ and weights w_2, w_3, w_4. Both segments form a C^1 curve if

$$\frac{w_1}{u_1 - u_0}\Delta\mathbf{b}_1 = \frac{w_3}{u_2 - u_1}\Delta\mathbf{b}_2. \tag{13.3}$$

The appearance of the interval lengths is due to the application of the chain rule, which is necessary since we now consider a composite curve with a global parameter u. Notice the absence of the weight w_2 in the C^1 equation!

13.4 The Circle

Of all conics, the circular arc is the one most widely used. Here we will represent it as a rational quadratic Bézier curve. Its control polygon must satisfy a special condition: It has to form an isosceles triangle, due to the circle's symmetry properties. Referring to Figure 13.4, and assuming standard form, we need to set

$$z_1 = \cos\alpha,$$

with $\alpha = \angle(\mathbf{b}_2, \mathbf{b}_0, \mathbf{b}_1)$.

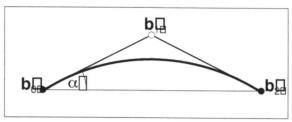

Figure 13.4.
The circle: the geometry of the control polygon.

A whole circle may be represented in many ways by piecewise rational quadratics. One example is to represent one quarter with the control polygon, and then use the complementary segment to write the remaining part. It is probably more convenient—retaining the convex hull property for positive weights—to dissect the full circle into four parts, as shown in Figure 13.5.

Although we can write an arc of a circle in rational quadratic form, one should not overlook that we then sacrifice one nice property of the familiar sin/cos parametrization: In the rational quadratic form, the parameter t does not traverse the circle with *unit speed*. Thus,

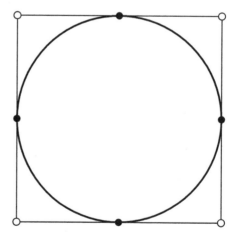

Figure 13.5.
The full circle: It may be represented by four rational quadratics.

if an arc of a rational quadratic is to be split into a certain number of segments, each subtending the same angle, numerical techniques must be invoked. In the sin/cos parametrization, by contrast, equal parameter increments ensure equal subtended angles.

13.5 Rational Bézier Curves

So far, we have obtained a conic section in $I\!E^2$ as the projection of a parabola (a quadratic) in $I\!E^3$. Conic sections may be expressed as rational quadratic Bézier curves, and their generalization to higher degree rational curves is quite straightforward: A rational Bézier curve of degree n in $I\!E^3$ is the projection of an nth degree Bézier curve in $I\!E^4$ into the hyperplane $w = 1$. Here, we denote 4D points by four coordinates and their 3D projections by three coordinates:

$$\underline{\mathbf{x}} = \begin{bmatrix} x \\ y \\ z \\ w \end{bmatrix} \longrightarrow \begin{bmatrix} x/w \\ y/w \\ z/w \end{bmatrix} = \mathbf{x}.$$

We may view this 4D hyperplane as a copy of $I\!E^3$; we assume that a point $\underline{\mathbf{x}}$ in $I\!E^4$ is given by four coordinates. Proceeding in exactly the same way as we did for conics, we can show that an nth degree rational Bézier curve is given by

$$\mathbf{x}(t) = \frac{w_0 \mathbf{b}_0 B_0^n(t) + \cdots + w_n \mathbf{b}_n B_n^n(t)}{w_0 B_0^n(t) + \cdots + w_n B_n^n(t)}; \quad \mathbf{x}(t), \mathbf{b}_i \in I\!E^3. \quad (13.4)$$

The w_i are again called *weights*; the $\mathbf{b}_i$ form the control polygon. It is the projection of the 4D control polygon $\underline{\mathbf{b}}_0, \ldots, \underline{\mathbf{b}}_n$. This 4D control polygon defines a 4D polynomial curve—the homogeneous form of the curve. It is given by

$$\underline{\mathbf{x}}(t) = \underline{\mathbf{b}}_0 B_0^n(t) + \cdots + \underline{\mathbf{b}}_n B_n^n(t).$$

In order to evaluate a rational Bézier curve, we apply the de Casteljau algorithm to this homogeneous form and project the resulting point into 3D.

EXAMPLE 13.3

We will evaluate the following Bézier curve $t = 0.5$. Take the control points from Example 3.3; They are

$$\begin{bmatrix} -1 \\ 0 \end{bmatrix}, \begin{bmatrix} 0 \\ 1 \end{bmatrix}, \begin{bmatrix} 0 \\ -1 \end{bmatrix}, \begin{bmatrix} 1 \\ 0 \end{bmatrix}.$$

However, make their weights $1, 2, 1, 1$. This gives the homogeneous control points

$$\begin{bmatrix} -1 \\ 0 \\ 1 \end{bmatrix}, \begin{bmatrix} 0 \\ 2 \\ 2 \end{bmatrix}, \begin{bmatrix} 0 \\ -1 \\ 1 \end{bmatrix}, \begin{bmatrix} 1 \\ 0 \\ 1 \end{bmatrix}.$$

Applying the de Casteljau algorithm to the homogeneous control points gives

$$\begin{bmatrix} -1.0 \\ 0.0 \\ 1 \end{bmatrix}$$
$$\begin{bmatrix} 0.0 \\ 2.0 \\ 2 \end{bmatrix} \quad \begin{bmatrix} -0.5 \\ 1.0 \\ 1.5 \end{bmatrix}$$
$$\begin{bmatrix} 0.0 \\ -1.0 \\ 1 \end{bmatrix} \quad \begin{bmatrix} 0.0 \\ 0.5 \\ 1.5 \end{bmatrix} \quad \begin{bmatrix} -0.25 \\ 0.75 \\ 1.5 \end{bmatrix}$$
$$\begin{bmatrix} 1.0 \\ 0.0 \\ 1.0 \end{bmatrix} \quad \begin{bmatrix} 0.5 \\ -0.5 \\ 1.0 \end{bmatrix} \quad \begin{bmatrix} 0.25 \\ 0 \\ 1.25 \end{bmatrix} \quad \begin{bmatrix} 0.0 \\ 0.375 \\ 1.375 \end{bmatrix}$$

thus the homogeneous form of the desired point is

$$\underline{\mathbf{x}}(0.5) = \begin{bmatrix} 0.0 \\ 0.375 \\ 1.375 \end{bmatrix}.$$

The corresponding 2D point is found after division by the third coordinate:

$$\mathbf{x}(0.5) = \begin{bmatrix} 0.0 \\ 0.2727 \end{bmatrix}.$$

Compare this example with Example 3.5!

If all weights are one, we obtain the standard nonrational Bézier curve; in that case, the denominator is identically equal to one. If some w_i are negative, singularities may occur; we will therefore only deal with nonnegative w_i. Rational Bézier curves enjoy all the properties that their nonrational counterparts possess; for example, they

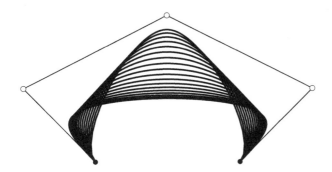

Figure 13.6.

Weights: As the weight of the Bézier point b_2 changes, the curves change as well.

are affinely invariant. If all w_i are nonnegative, we have the convex hull property.[2]

The influence of the weights is illustrated in Figure 13.6. The "top" curve corresponds to $w_2 = 10$; the "bottom" one corresponds to $w_2 = 0.1$.

Rational Bézier curves enjoy a property which is not shared by their nonrational brethren: This is *projective invariance*. A projective map maps homogeneous coordinates $\underline{\mathbf{x}}$ to new homogeneous coordinates $\underline{\bar{\mathbf{x}}}$. It takes the form of a linear map

$$\underline{\bar{\mathbf{x}}} = A\underline{\mathbf{x}}$$

with A being a 4×4 matrix. Such a map will change the weights of a curve. For the simple example of rational quadratic conics, projective maps are capable of mapping an ellipse to a hyperbola!

The curvature and torsion formulas from Section 8.2 change just slightly for rational curves. At $t = 0$ we have

$$\kappa(0) = 2\frac{n-1}{n}\frac{w_0 w_2}{w_1}\frac{\text{area}[\mathbf{b}_0, \mathbf{b}_1, \mathbf{b}_2]}{\|\mathbf{b}_1 - \mathbf{b}_0\|^3}$$

and

$$\tau(0) = \frac{3}{2}\frac{n-2}{n}\frac{w_0 w_3}{w_1 w_2}\frac{\text{volume}[\mathbf{b}_0, \mathbf{b}_1, \mathbf{b}_2, \mathbf{b}_3]}{\text{area}[\mathbf{b}_0, \mathbf{b}_1, \mathbf{b}_2]^2}.$$

---

[2]From Figure 13.3 it should be clear that weights with changing signs do not produce this property.

13.6   Rational B-Spline Curves

Rational B-spline curves, known as NURBS, short for *NonUniform Rational B-spline curveS* have become a standard in the CAD/CAM industry.[3] They are defined in a not too surprising fashion:

$$\mathbf{x}(u) = \frac{w_0 \mathbf{d}_0 N_0^n(u) + \ldots + w_{D-1} \mathbf{d}_{D-1} N_{D-1}^n(u)}{w_0 N_0^n(u) + \ldots + w_{D-1} N_{D-1}^n(u)}. \tag{13.5}$$

All properties from the rational Bézier form carry over, such as convex hull (for nonnegative weights), or affine and projective invariance.

Derivatives may easily be computed using the equations of Section 13.3.

Designing with cubic NURB curves is not very different from designing with their nonrational counterparts. But we now have the added freedom of being able to change weights. A change of only one weight affects a rational B-spline curve only locally.

13.7   Rational Bézier and B-Spline Surfaces

We can generalize Bézier and B-spline surfaces to their rational counterparts in much the same way as we did for the curve cases. In other words, we define a rational Bézier or B-spline surface as the projection of a 4D tensor product Bézier or B-spline surface. Thus, the rational Bézier patch takes the form

$$\mathbf{x}(u, v) = \frac{M^{\mathrm{T}} \mathbf{B}_w N}{M^{\mathrm{T}} W N}. \tag{13.6}$$

The notation is that of (6.8), but now the matrix $\mathbf{B}_w$ has elements $w_{i,j} \mathbf{b}_{i,j}$ and the matrix W has elements $w_{i,j}$. These $w_{i,j}$ are again called weights and influence the shape of the surface in much the same way as we observed for the curve case.

A rational B-spline surface is similarly written as

$$\mathbf{s}(u, v) = \frac{M^{\mathrm{T}} \mathbf{D}_w N}{M^{\mathrm{T}} W N}, \tag{13.7}$$

where the matrices M and N contain the B-spline basis functions in u and v.

---

[3]This is an unfortunate misnomer. Using the strict definition, *uniform* B-spline curves do not qualify as NURBS, although they are used frequently and, in fact, fit the standard data exchange formats for NURBS.

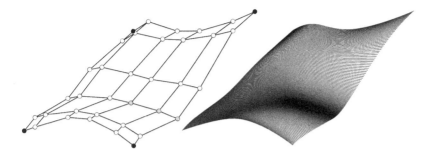

Figure 13.7.
A NURB surface.

Figure 13.7 shows a rational B-spline surface. Its control net is identical to the one from Figure 12.6. However, the weights of the gray control points are now set to 3. The resulting surface, while only changed locally, draws closer to the control points with increased weights. This is in complete analogy to the curve case.

13.8   Surfaces of Revolution

One advantage of rational B-spline surfaces is that they allow the exact representation of surfaces of revolution. A surface of revolution is obtained by rotating or sweeping a curve—the *generatrix*—around an axis. Our generatrix will be of the form

$$\mathbf{g}(v) = \begin{bmatrix} r(v) \\ 0 \\ z(v) \end{bmatrix}.$$

This planar curve in the (x, z)-plane would be a (rational) Bézier curve or a B-spline curve in most practical cases. For our axis of revolution, we will take the z-axis. In Figure 13.8, the z-axis comes out of the center of the half-torus. The (x, z)-plane is nearly aligned with the view.

In this context, a surface of revolution is given by

$$\mathbf{x}(u, v) = \begin{bmatrix} r(v) \cos u \\ r(v) \sin u \\ z(v) \end{bmatrix}$$

For fixed v, an isoparametric line $v = const$ traces out a circle of radius $r(v)$, called a *meridian*. Since a circle may be exactly represented by rational quadratic arcs, we may find an exact rational representation of a surface of revolution given that $r(v)$ and $z(v)$ are in rational form.

Let

$$\mathbf{c}_i = \begin{bmatrix} x_i \\ 0 \\ z_i \end{bmatrix}$$

be the control points of the generatrix and let w_i be their weights.[4] Then, the surface of revolution is broken down into four symmetric pieces which are rational quadratic in the parameter u. Each piece corresponds to one quadrant of the (x, y)-plane.

Over the first quadrant, we have a surface with three columns of control points and associated weights. They are given by

$$\begin{bmatrix} x_i \\ 0 \\ z_i \end{bmatrix}, \quad \begin{bmatrix} x_i \\ x_i \\ z_i \end{bmatrix}, \quad \begin{bmatrix} 0 \\ x_i \\ z_i \end{bmatrix}.$$

Their weights are $w_i, \frac{\sqrt{2}}{2}w_i, w_i$. The remaining three surface segments are now simply obtained by reflecting this one appropriately.

We finish this section with an example of a NURB surface of revolution.

EXAMPLE 13.4

A *torus* is a surface of revolution. We consider one-sixteenth of a torus that has been created by revolving a quarter circle, defined in the (x, z)-plane and centered at $\begin{bmatrix} 2 \\ 0 \\ 0 \end{bmatrix}$ around the z-axis. This quarter circle, or generatrix of the torus piece, is given by Bézier points

$$\begin{bmatrix} 2 \\ 0 \\ 1 \end{bmatrix} \begin{bmatrix} 3 \\ 0 \\ 1 \end{bmatrix} \begin{bmatrix} 3 \\ 0 \\ 0 \end{bmatrix}$$

and weights $1, \sqrt{2}/2, 1$.

---

[4]It does not matter if the $\mathbf{c}_i$ are Bézier or B-spline control points.

The control points for a rational biquadratic patch on this one-sixteenth of the torus:

$$\begin{bmatrix} 2 \\ 0 \\ 1 \end{bmatrix} \begin{bmatrix} 3 \\ 0 \\ 1 \end{bmatrix} \begin{bmatrix} 3 \\ 0 \\ 0 \end{bmatrix}$$
$$\begin{bmatrix} 2 \\ 2 \\ 1 \end{bmatrix} \begin{bmatrix} 3 \\ 3 \\ 1 \end{bmatrix} \begin{bmatrix} 3 \\ 3 \\ 0 \end{bmatrix}$$
$$\begin{bmatrix} 0 \\ 2 \\ 1 \end{bmatrix} \begin{bmatrix} 0 \\ 3 \\ 1 \end{bmatrix} \begin{bmatrix} 0 \\ 3 \\ 0 \end{bmatrix}$$

with weights

$$\begin{bmatrix} 1 & \frac{\sqrt{2}}{2} & 1 \\ \frac{\sqrt{2}}{2} & \frac{1}{2} & \frac{\sqrt{2}}{2} \\ 1 & \frac{\sqrt{2}}{2} & 1 \end{bmatrix}$$

In Figure 13.8, we use two rational Bézier quadratics to generate the top of a torus. The quarter circle in the Example above is extended to the half circle with control points

$$\begin{bmatrix} 1 \\ 0 \\ 0 \end{bmatrix} \begin{bmatrix} 1 \\ 0 \\ 1 \end{bmatrix} \begin{bmatrix} 2 \\ 0 \\ 1 \end{bmatrix} \begin{bmatrix} 3 \\ 0 \\ 1 \end{bmatrix} \begin{bmatrix} 3 \\ 0 \\ 0 \end{bmatrix},$$

and weights $1, \sqrt{2}/2, 1, \sqrt{2}/2, 1$.

13.9 Exercises

1. Let a conic be given by control points

$$\begin{bmatrix} -1 \\ 0 \end{bmatrix}, \begin{bmatrix} 0 \\ 1 \end{bmatrix}, \begin{bmatrix} 2 \\ 0 \end{bmatrix},$$

and weights $2, 2, 4$. What type of conic is this?

2. Repeat the Exercise 1 with weights $1, 2, 4$.

3. Redo Example 13.3 for the same control polygon but for weights $1, 2, 2, 2$.

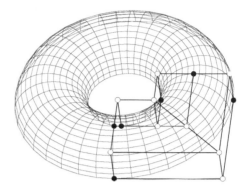

Figure 13.8.
Surfaces of revolution: the top of a torus is represented by eight rational biquadratic patches. The control nets of two of them are shown.

4. Let a quadrant of a circle $\mathbf{x}(t)$ be given by control points

$$\begin{bmatrix} 2 \\ 0 \end{bmatrix}, \begin{bmatrix} 2 \\ 2 \end{bmatrix}, \begin{bmatrix} 0 \\ 2 \end{bmatrix}$$

and weights $1, \sqrt{2}/2, 1$. Compute $\|\dot{\mathbf{x}}(0)\|$ and $\|\dot{\mathbf{x}}(0.5)\|$.

5. Let a rational cubic generatrix be given by control points

$$\begin{bmatrix} 2 \\ 0 \\ 0 \end{bmatrix} \begin{bmatrix} 2 \\ 0 \\ 2 \end{bmatrix} \begin{bmatrix} 1 \\ 0 \\ 1 \end{bmatrix} \begin{bmatrix} 2 \\ 0 \\ 4 \end{bmatrix}$$

and weights $1, 2, 2, 1$. What are the Bézier points of the first quadrant of the corresponding surface of revolution?

6. Let a generatrix be given by the straight line segment through

$$\begin{bmatrix} 2 \\ 0 \\ 0 \end{bmatrix} \quad \text{and} \quad \begin{bmatrix} 0 \\ 0 \\ 2 \end{bmatrix}.$$

As we rotate it around the z-axis, we obtain a cone. What are the Bézier points of its first quadrant?

Hunting Geometry Bugs

14

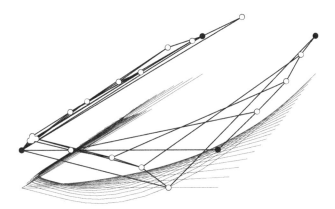

Figure 14.1.
A bug in a program.

When you code up any of the concepts in this book, chances are that your program doesn't work right away—for example, we produced Figure 14.1 in an early attempt for one of the figures in the book. Here are some hints on how to fix some common errors.

Equality check: A typical error that inexperienced programmers will almost invariably produce will look like this:

```
if (x == y) return;
```

What is bad about this? We assume that x and y are reals, i.e., represented as `float` or `double`. Inside a larger program, they will both be the result of some computation. Then one can almost guarantee that they will never be equal! Keep in mind that in order for two reals to be equal, *all digits* must be equal. But computation produces roundoff, and so it is more than likely that the last digits will differ.

The safest way out of this problem is to employ a tolerance `tol`. Then the above would look like this:

$$\texttt{if (fabs(x - y) < tol) return;}$$

The value of `tol` depends on the application at hand and may critically influence your computations. Without any knowledge about an application, `tol=1.0E-6` should work.

In the same way, it is also not safe to check for numbers being positive or negative: `if (x < 0)` should be replaced by `if x < -tol`.

Test case size: More often than not, novice programmers try out their code on real data sets, often involving hundreds or thousands of input numbers. The resulting printouts become very tedious to read! Create a simple, even trivial test case. Debug it thoroughly, and then move on to larger data sets.

Smart test cases: In many curve or surface algorithms, linear input data result in linear output. For example, if all control points of a Bézier or B-spline curve are collinear, then the resulting curve will be a straight line. Try out your program on linear data sets; if straight lines (or planes) are not reproduced, this might give you a clue as to where to look for bugs.

If you work with surface algorithms, try to work with simple examples, gradually increasing their complexity. Thus, you might want to check your code for a sequence of surfaces such as $z = 0, z = 1, z = x, z = 2x, z = x^2, z = x^2 + y^2$, etc. For these surfaces (in parametric form), you might know what your code should produce and thus you have an array of test cases.

Scale and translation invariance: If you write a piece of geometry code for some application, it should not matter if you solve your problem for a data set at a particular location or if the data set is translated to a different location. If a piece of code acts up on you, one quick debugging trick is to check if it is invariant under translations: Run your code on a simple data set, then run it on the same set, but translated by some amount. If you get different results for the original and the translated data, you should have a clue as to where to look for bugs.

Barycentric combinations: Once you established that your program is not affinely invariant, a likely source for this is the use of nonbarycentric combinations. If a point is computed as the result of a linear combination of other points, then that linear combination must be *barycentric*—the coefficients must sum to one.

Solutions to Selected Exercises

A

Chapter 1. The Bare Basics

3. The desired ratio is $1/3$.

5. The barycentric coordinates of the points are as follows.

$$\mathbf{p}_1 \cong (\frac{1}{3}, \frac{1}{3}, \frac{1}{3}) \quad \mathbf{p}_2 \cong (\frac{1}{2}, \frac{1}{2}, 0) \quad \mathbf{p}_3 \cong (0, 0, 1)$$

6. The triangle area is $9/2$.

Chapter 2. Lines and Planes

2. The explicit form is given by $y = -x + 1$.

4. The explicit form is given by $z = x + y$. Hence, the implicit form is $x + y - z = 0$.

7. If there are no "-1" entries, every edge is an interior edge. Thus the corresponding mesh is *closed*. An example would be four triangles forming a tetrahedron.

Chapter 3. Cubic Bézier Curves

1. If all Bézier points are the same point, then the whole curve is confined to that point.

3. The desired sketch is Sketch 104. The point on the curve is $\begin{bmatrix} -22/32 \\ 36/64 \end{bmatrix}$ and the derivative is $3 \times \begin{bmatrix} 6/8 \\ 8/16 \end{bmatrix}$.

5. The curve $[t, B_0^3]^{\mathrm{T}}$ is given by the Bézier points

$$\begin{bmatrix} 0 \\ 1 \end{bmatrix}, \quad \begin{bmatrix} \frac{1}{3} \\ 0 \end{bmatrix}, \quad \begin{bmatrix} \frac{2}{3} \\ 0 \end{bmatrix}, \quad \begin{bmatrix} 1 \\ 0 \end{bmatrix}.$$

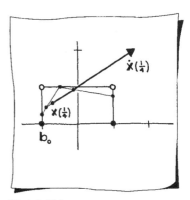

Sketch 104.
Solution to Exercise 3 in Section 3.8.

211

The curve $[t, B_1^3]^T$ is given by the Bézier points

$$\begin{bmatrix} 0 \\ 0 \end{bmatrix}, \quad \begin{bmatrix} \frac{1}{3} \\ 1 \end{bmatrix}, \quad \begin{bmatrix} \frac{2}{3} \\ 0 \end{bmatrix}, \quad \begin{bmatrix} 1 \\ 0 \end{bmatrix}.$$

The other two follow similarly.

7. The desired Bézier points are given by

$$\begin{bmatrix} 0 \\ 0 \\ 0 \end{bmatrix}, \quad \begin{bmatrix} \frac{1}{3} \\ 0 \\ 0 \end{bmatrix}, \quad \begin{bmatrix} \frac{2}{3} \\ \frac{1}{3} \\ 0 \end{bmatrix}, \quad \begin{bmatrix} 1 \\ 1 \\ 1 \end{bmatrix}.$$

Chapter 4. Bézier Curves: Cubic and Beyond

1. The solution is shown in Figure A.1

3. The computation count depends on how the algorithm is organized. Just counting the elements in the triangular de Casteljau

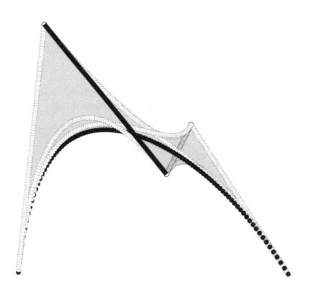

Figure A.1.
The trajectory of the points $\mathbf{b}_1^1(t)$ together with the points $\mathbf{b}_0^4(t)$

scheme gives an $O(n^2)$. But the count is much higher if one uses a recursive call like

```
decas(n,b[0...n],t)=
(1-t)*decas(n-1,b[0...n-1],t) + t*decas(n-1,b[1...n],t);
```

6. Solving (4.22) results in the following quadratic curve

$$\mathbf{b}_0 = \begin{bmatrix} -3.6 \\ 1.2 \\ 0 \end{bmatrix}, \quad \mathbf{b}_1 = \begin{bmatrix} 0 \\ 0 \\ 0 \end{bmatrix}, \quad \mathbf{b}_2 = \begin{bmatrix} 3.6 \\ -1.2 \\ 0 \end{bmatrix}.$$

Forcing endpoint interpolation changes the curve to

$$\mathbf{b}_0 = \begin{bmatrix} -4 \\ 0 \\ 0 \end{bmatrix}, \quad \mathbf{b}_1 = \begin{bmatrix} 0 \\ 0 \\ 0 \end{bmatrix}, \quad \mathbf{b}_2 = \begin{bmatrix} 4 \\ 0 \\ 0 \end{bmatrix}.$$

7. Over the interval $[0, 1]$, the Bézier points are

$$\mathbf{b}_0 = \begin{bmatrix} 0 \\ 0 \end{bmatrix}, \quad \mathbf{b}_1 = \begin{bmatrix} \frac{1}{2} \\ 1 \end{bmatrix}, \quad \mathbf{b}_2 = \begin{bmatrix} 1 \\ 3 \end{bmatrix}.$$

Over the interval $[-1, 1]$, the Bézier points are

$$\mathbf{b}_0 = \begin{bmatrix} -1 \\ -1 \end{bmatrix}, \quad \mathbf{b}_1 = \begin{bmatrix} 0 \\ -1 \end{bmatrix}, \quad \mathbf{b}_2 = \begin{bmatrix} 1 \\ 3 \end{bmatrix}.$$

Chapter 5. Putting Curves to Work

1. The cubic Bézier control polygon is given by

$$\mathbf{p}_0 = \begin{bmatrix} 0 \\ 0 \end{bmatrix}, \quad \mathbf{p}_1 = \begin{bmatrix} 1 \\ 0 \end{bmatrix}, \quad \mathbf{p}_2 = \begin{bmatrix} 2 \\ 3 \end{bmatrix}, \quad \mathbf{p}_3 = \begin{bmatrix} 3 \\ 9 \end{bmatrix}.$$

4. The three Lagrange polynomials are shown in Sketch 105. They include:

$$L_0^2(t) = \frac{(t-4)(t-5)}{20} \quad L_1^2(t) = \frac{t(t-5)}{-4} \quad L_2^2(t) = \frac{t(t-4)}{5}$$

5. For the first one, we that have $\mathbf{x}(0), \dot{\mathbf{x}}(0), \dot{\mathbf{x}}(1), \mathbf{x}(1)$ are given by

$$\begin{bmatrix} 0 \\ 0 \end{bmatrix}, \quad \begin{bmatrix} 4.5 \\ 3 \end{bmatrix}, \quad \begin{bmatrix} 4.5 \\ -3 \end{bmatrix}, \quad \begin{bmatrix} 1 \\ 0 \end{bmatrix}.$$

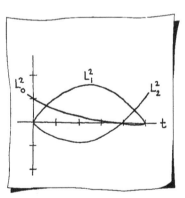

Sketch 105.
Solution to Exercise 4.

Chapter 6. Bézier Patches

2. The desired mixed partial is the zero vector:

$$\mathbf{x}_{uv}(0.5, 0.5) = \begin{bmatrix} 0 \\ 0 \\ 0 \end{bmatrix}.$$

It is easily obtained as the twist vector of the bilinear patch which was computed in Example 6.9 via the 3-stage de Casteljau evaluation method.

The closed-form expression is

$$\mathbf{x}_{uv}(0.5, 0.5) = \begin{bmatrix} \frac{1}{2} & \frac{1}{2} \end{bmatrix} \begin{bmatrix} \begin{bmatrix} 0 \\ 0 \\ 3 \end{bmatrix} & \begin{bmatrix} 0 \\ 0 \\ 0 \end{bmatrix} & \begin{bmatrix} 0 \\ 0 \\ -6 \end{bmatrix} \\ \begin{bmatrix} 0 \\ 0 \\ -3 \end{bmatrix} & \begin{bmatrix} 0 \\ 0 \\ 0 \end{bmatrix} & \begin{bmatrix} 0 \\ 0 \\ 6 \end{bmatrix} \end{bmatrix} \begin{bmatrix} \frac{1}{4} \\ \frac{1}{2} \\ \frac{1}{4} \end{bmatrix}.$$

3. The new control net is given by

$$\begin{bmatrix} \begin{bmatrix} 0 \\ 0 \\ 6 \end{bmatrix} & \begin{bmatrix} 3 \\ 0 \\ 0 \end{bmatrix} & \begin{bmatrix} 6 \\ 0 \\ 0 \end{bmatrix} & \begin{bmatrix} 9 \\ 0 \\ 6 \end{bmatrix} \\ \begin{bmatrix} 0 \\ 2 \\ 4 \end{bmatrix} & \begin{bmatrix} 3 \\ 2 \\ 0 \end{bmatrix} & \begin{bmatrix} 6 \\ 2 \\ 0 \end{bmatrix} & \begin{bmatrix} 9 \\ 2 \\ 2 \end{bmatrix} \\ \begin{bmatrix} 0 \\ 4 \\ 4 \end{bmatrix} & \begin{bmatrix} 3 \\ 4 \\ 0 \end{bmatrix} & \begin{bmatrix} 6 \\ 4 \\ 0 \end{bmatrix} & \begin{bmatrix} 9 \\ 4 \\ 2 \end{bmatrix} \\ \begin{bmatrix} 0 \\ 6 \\ 6 \end{bmatrix} & \begin{bmatrix} 3 \\ 6 \\ 0 \end{bmatrix} & \begin{bmatrix} 6 \\ 6 \\ 0 \end{bmatrix} & \begin{bmatrix} 9 \\ 6 \\ 6 \end{bmatrix} \end{bmatrix}$$

8. Over the unit square, the control points are

$$
\left[
\begin{array}{ccc}
\left[\begin{array}{c} 0 \\ 0 \\ 0 \end{array}\right] &
\left[\begin{array}{c} \frac{1}{2} \\ 0 \\ 0 \end{array}\right] &
\left[\begin{array}{c} 1 \\ 0 \\ 1 \end{array}\right] \\
\left[\begin{array}{c} 0 \\ \frac{1}{2} \\ 0 \end{array}\right] &
\left[\begin{array}{c} \frac{1}{2} \\ \frac{1}{2} \\ 0 \end{array}\right] &
\left[\begin{array}{c} 1 \\ \frac{1}{2} \\ 1 \end{array}\right] \\
\left[\begin{array}{c} 0 \\ 1 \\ 0 \end{array}\right] &
\left[\begin{array}{c} \frac{1}{2} \\ 1 \\ 0 \end{array}\right] &
\left[\begin{array}{c} 1 \\ 1 \\ 1 \end{array}\right]
\end{array}
\right]
$$

Hint: Use derivatives to find the control points for the patch over $[1,1] \times [3,3]$.

10. All you need to do is to degree elevate curve $\mathbf{c}$. Then the Bézier points of $\mathbf{c}$ and $\mathbf{d}$ form a ruled surface of degrees 3×1.

Chapter 7. Working with Polynomial Patches

3. The monomial form of the interpolant is given by

$$
\left[\begin{array}{cc} 1 & u \end{array}\right]
\left[
\begin{array}{cc}
\left[\begin{array}{c} 0 \\ 0 \\ 0 \end{array}\right] &
\left[\begin{array}{c} 1 \\ 0 \\ 0 \end{array}\right] \\
\left[\begin{array}{c} 0 \\ 2 \\ 1 \end{array}\right] &
\left[\begin{array}{c} 0 \\ 0 \\ -2 \end{array}\right]
\end{array}
\right]
\left[\begin{array}{c} 1 \\ v \end{array}\right]
$$

4. The missing control point is

$$
\mathbf{b}_{1,1} = \left[\begin{array}{c} 0.5 \\ 0.5 \\ 0.5 \end{array}\right].
$$

8. The bilinear surface now has the control polygon

$$
\left[
\begin{array}{cc}
\left[\begin{array}{c} -2 \\ -2 \\ 1 \end{array}\right] &
\left[\begin{array}{c} 1 \\ 0 \\ 0 \end{array}\right] \\
\left[\begin{array}{c} 0 \\ 1 \\ 0 \end{array}\right] &
\left[\begin{array}{c} 2 \\ 2 \\ 1 \end{array}\right]
\end{array}
\right].
$$

Reason: the new fifth point lies on the bilinear patch through the first four points.

Chapter 8. Shape

2. The desired curvatures are given by

$$\kappa(0) = \frac{2}{3} \qquad \kappa(1) = \frac{1}{3}.$$

4. At $\mathbf{x}(0,0)$, we have $M = 0$.

6. The Gaussian curvature decreases by a factor of 4 (the new surface is flatter than the original one); the mean curvature remains unchanged.

Chapter 9. Composite Curves

2. The composite curve is C^1 if the knot sequence is given $0, \alpha, 1$ or some multiple thereof. It is then also C^2. Irrespective of a knot sequence, the composite curve is G^1 and G^2.

4. The junction point is

$$\mathbf{b}_3 = \left[\begin{array}{c} 3 \\ 2 \end{array} \right].$$

6. A possible knot sequence is given by $[0, 1, 3]$. Any multiple will also qualify.

9. The projection is the point

$$\left[\begin{array}{c} 2 \\ 0 \\ 1 \end{array} \right].$$

Chapter 10. B-Spline Curves

2. The multiplicity vector is given by

$$2, 0, 1, 2, 0, 1, 2, 0.$$

A quadratic B-spline curve will have four segments and a cubic will have two segments.

4. The derivative is

$$\dot{\mathbf{x}}(0) = \left[\begin{array}{c} 3 \\ 3 \end{array} \right].$$

6. The answer is provided by Sketch 106.

8. The cubic control points are given by

$$\begin{bmatrix} 0 \\ 0 \end{bmatrix}, \begin{bmatrix} 1 \\ 3 \end{bmatrix}, \begin{bmatrix} \frac{7}{3} \\ 7 \end{bmatrix}, \begin{bmatrix} \frac{13}{3} \\ 13 \end{bmatrix}, \begin{bmatrix} \frac{16}{3} \\ 16 \end{bmatrix}, \begin{bmatrix} 6 \\ 18 \end{bmatrix}.$$

10. The knot vector must have full multiplicity at each knot.

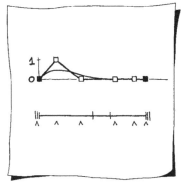

Sketch 106.
The cubic B-spline for Exercise 6.

Chapter 11. Working with B-Spline Curves

2. A chord length knot sequence would be

$$0, 2, 4, 5, 7.236.$$

Any multiple will also qualify.

4. The Bessel tangents are

$$\mathbf{t}_s = \begin{bmatrix} 3 \\ 2 \end{bmatrix} \qquad \mathbf{t}_e = \begin{bmatrix} 3 \\ -2 \end{bmatrix}$$

5. The system of equations takes the form

$$\begin{bmatrix} 1 & 0 & 0 \\ 1 & 4 & 1 \\ 0 & 0 & 1 \end{bmatrix} \begin{bmatrix} \mathbf{d}_0 \\ \mathbf{d}_1 \\ \mathbf{d}_2 \end{bmatrix} = \begin{bmatrix} \begin{bmatrix} 1 \\ \frac{2}{3} \end{bmatrix} \\ \begin{bmatrix} 18 \\ 6 \end{bmatrix} \\ \begin{bmatrix} 5 \\ \frac{2}{3} \end{bmatrix} \end{bmatrix}.$$

The cubic spline interpolant over a uniform knot sequence has control points

$$\mathbf{d}_0 = \begin{bmatrix} 0 \\ 0 \end{bmatrix}, \mathbf{d}_1 = \begin{bmatrix} 1 \\ \frac{2}{3} \end{bmatrix}, \mathbf{d}_2 = \begin{bmatrix} 3 \\ \frac{7}{6} \end{bmatrix}, \mathbf{d}_3 = \begin{bmatrix} 5 \\ \frac{2}{3} \end{bmatrix}, \mathbf{d}_4 = \begin{bmatrix} 6 \\ 0 \end{bmatrix}.$$

Chapter 12. Composite Surfaces

1. For the "left" surface, we have

$$\mathbf{x}_u(1, 0.5) = \begin{bmatrix} 1 \\ 0 \\ 0.5 \end{bmatrix}.$$

For the "right" surface, we have

$$\mathbf{x}_u(1, 0.5) = \begin{bmatrix} 2 \\ 0 \\ 0.5 \end{bmatrix}.$$

These vectors are not equal, hence the surface is not C^1.

3. The control net is obtained through

$$\mathbf{c}_{i,j} = \begin{bmatrix} \begin{bmatrix} 0 \\ 0 \\ 0 \end{bmatrix} & \begin{bmatrix} 2 \\ 0 \\ 0 \end{bmatrix} & \begin{bmatrix} 4 \\ 0 \\ 0 \end{bmatrix} & \begin{bmatrix} 6 \\ 0 \\ 0 \end{bmatrix} \\ \begin{bmatrix} 0 \\ 6 \\ 0 \end{bmatrix} & \begin{bmatrix} 2 \\ 6 \\ 2 \end{bmatrix} & \begin{bmatrix} 4 \\ 6 \\ 4 \end{bmatrix} & \begin{bmatrix} 6 \\ 6 \\ 6 \end{bmatrix} \end{bmatrix}.$$

Hence

$$\mathbf{d}_{i,j} = \begin{bmatrix} \begin{bmatrix} 0 \\ 0 \\ 0 \end{bmatrix} & \begin{bmatrix} 2 \\ 0 \\ 0 \end{bmatrix} & \begin{bmatrix} 4 \\ 0 \\ 0 \end{bmatrix} & \begin{bmatrix} 6 \\ 0 \\ 0 \end{bmatrix} \\ \begin{bmatrix} 0 \\ 2 \\ 0 \end{bmatrix} & \begin{bmatrix} 2 \\ 2 \\ 2/3 \end{bmatrix} & \begin{bmatrix} 4 \\ 2 \\ 4/3 \end{bmatrix} & \begin{bmatrix} 6 \\ 2 \\ 2 \end{bmatrix} \\ \begin{bmatrix} 0 \\ 4 \\ 4/3 \end{bmatrix} & \begin{bmatrix} 2 \\ 4 \\ 8/3 \end{bmatrix} & \begin{bmatrix} 4 \\ 4 \\ 4 \end{bmatrix} & \begin{bmatrix} 6 \\ 4 \\ 4 \end{bmatrix} \\ \begin{bmatrix} 0 \\ 6 \\ 0 \end{bmatrix} & \begin{bmatrix} 2 \\ 6 \\ 4/3 \end{bmatrix} & \begin{bmatrix} 4 \\ 6 \\ 8/3 \end{bmatrix} & \begin{bmatrix} 6 \\ 6 \\ 4 \end{bmatrix} \end{bmatrix}.$$

5. The solution is

$$\mathbf{b}_{0,2} = \begin{bmatrix} 0 \\ 12 \\ -6 \end{bmatrix}, \mathbf{b}_{1,2} = \begin{bmatrix} 10 \\ 18 \\ 6 \end{bmatrix}$$

Chapter 13. NURBS

1. In order to classify the conic, we need to bring it into standard form. We first divide by w_0 and obtain weights $1, 1, 2$. The reparametrization factor is $c = 1/\sqrt{2}$, giving standard wights $1, \frac{1}{\sqrt{2}}, 1$ Thus the conic is a hyperbola.

3. Now the homogeneous control points are

$$\begin{bmatrix} -1 \\ 0 \\ 1 \end{bmatrix}, \begin{bmatrix} 0 \\ 2 \\ 2 \end{bmatrix}, \begin{bmatrix} 0 \\ -2 \\ 2 \end{bmatrix}, \begin{bmatrix} 2 \\ 0 \\ 2 \end{bmatrix}.$$

The homogeneous form of the desired point is

$$\underline{\mathbf{x}}(0.5) = \begin{bmatrix} 0.125 \\ 0 \\ 1.875 \end{bmatrix}.$$

6. The desired control net is given by

$$\begin{bmatrix} \begin{bmatrix} 2 \\ 0 \\ 0 \end{bmatrix} & \begin{bmatrix} 0 \\ 0 \\ 2 \end{bmatrix} \\ \begin{bmatrix} 2 \\ 2 \\ 0 \end{bmatrix} & \begin{bmatrix} 0 \\ 0 \\ 2 \end{bmatrix} \\ \begin{bmatrix} 0 \\ 2 \\ 0 \end{bmatrix} & \begin{bmatrix} 0 \\ 0 \\ 2 \end{bmatrix} \end{bmatrix}$$

with weights

$$\begin{bmatrix} 1 & 1 \\ \frac{\sqrt{2}}{2} & \frac{\sqrt{2}}{2} \\ 1 & 1 \end{bmatrix}.$$

This patch is degenerate at the top, as is necessary for a cone.

Notation B

Here is the notation used in this book:

$\wedge$	cross product
$\dot{}\,\ddot{}$	curve derivatives with respect to the current parameter
a, b, α, β	real numbers or real-valued functions
$\mathbf{0}$	short for $\begin{bmatrix} 0 \\ 0 \end{bmatrix}$ or $\begin{bmatrix} 0 \\ 0 \\ 0 \end{bmatrix}$
$\mathbf{e}_1, \mathbf{e}_2, \mathbf{e}_3$	directions of a coordinate system
$\mathbf{a}, \mathbf{b}$	points or vectors
a_x, a_y, a_z	x, y, z components of the point $\mathbf{a}$.
$\underline{\mathbf{x}}$	homogeneous coordinates of the point $\mathbf{x}$
A, B	matrices
$\mathbf{A}, \mathbf{B}$	matrices whose elements are points ("hypermatrices")
$\mathbf{b}_i^r$	intermediate points in the de Casteljau algorithm
B_i^n	Bernstein polynomial of degree n
H_i^3	cubic Hermite polynomials
L_i^n	Lagrange polynomial of degree n
N_i^n	B-spline basis function of degree n
$\mathbb{E}^d$	$d-$dimensional Euclidean or affine space
$\mathbb{R}^d$	$d-$dimensional linear or real space
Δ_i	difference in parameter intervals (i.e., $\Delta_i = u_{i+1} - u_i$)
Δ^r	iterated forward difference
$\mathbf{P}$	control polygon
$\|\mathbf{v}\|$	(Euclidean) length of the vector $\mathbf{v}$
$\mathbf{x}_u$	u-partial of $\mathbf{x}(u, v)$
$\mathbf{a} \cong (1, 0, 0)$	barycentric coordinates of $\mathbf{a}$ are $(1, 0, 0)$
ξ_i	Greville abscissa

Bibliography

[1] E. Angel. *Interactive Computer Graphics*. Addison-Wesley, 2000. Second edition.

[2] H. Anton. *Elementary Linear Algebra*. New York: John Wiley & Sons, 1981. Third edition.

[3] W. Boehm, G. Farin, and J. Kahmann. A survey of curve and surface methods in CAGD. *Computer Aided Geometric Design*, 1(1):1–60, 1984.

[4] W. Boehm and H. Prautzsch. *Geometric Concepts for Geometric Design*. Wellesley, MA: A K Peters Ltd., 1992.

[5] W. Boehm and H. Prautzsch. *Numerical Methods*. Vieweg, 1992.

[6] C. de Boor. *A Practical Guide to Splines*. Springer, 1978.

[7] M. do Carmo. *Differential Geometry of Curves and Surfaces*. Prentice Hall, Englewood Cliffs, 1976.

[8] G. Farin. *NURB Curves and Surfaces*. Wellesley, MA: A K Peters Ltd., 1995. Second edition 1999.

[9] G. Farin. *Curves and Surfaces for Computer Aided Geometric Design*. Boston: Academic Press, 1996. Fourth edition.

[10] G. Farin and D. Hansford. *Practical Linear Algebra: The Geometry Toolbox*. Wellesley, MA: A K Peters Ltd., 2005.

[11] I. Faux and M. Pratt. *Computational Geometry for Design and Manufacture*. West Sussex, England: Ellis Horwood Ltd., 1979.

[12] D. Hearn and M. Baker. *Computer Graphics*. Englewood Cliffs, NJ: Prentice-Hall, 1986.

[13] D. Hilbert and S. Cohn-Vossen. *Geometry and the Imagination*. Chelsea, New York, 1952.

[14] F. Hill. *Computer Graphics*. New York: Macmillan Publishing Company, 1990.

[15] J. Hoschek and D. Lasser. *Grundlagen der Geometrischen Datenverarbeitung*. Stuttgart: B.G. Teubner, 1989. English translation: *Fundamentals of Computer Aided Geometric Design*, Wellesley, MA: A K Peters Ltd., 1993.

[16] Adobe Systems Inc. *PostScript Language Tutorial and Cookbook*. Reading, MA: Addison-Wesley Publishing Company, Inc., 1985.

[17] L. Johnson and R. Riess. *Numerical Analysis*. Reading, MA: Addison-Wesley Publishing Company, Inc., 1982. Second edition.

[18] R. Liming. *Practical Analytical Geometry with Applications to Aircraft*. Macmillan, 1944.

[19] D. Marsh. *Applied Geometry for Computer Graphics and CAD*. Springer-Verlag, 1999.

[20] L. Piegl and W. Tiller. *The book of NURBS*. Springer-Verlag, 1995.

[21] A. Rockwood and P. Chambers. *Interactive Curves and Surfaces*. Morgan Kaufmann, 1996.

Index

absolute curvature, 124
affine invariance, 150
affine map, 7, 29, 78, 131
affine space, 2
Aitken's algorithm, 63
approximation, 52, 66
    least squares, 66, 106
area, 9
    3D triangle, 9
    parallelogram, 6
    signed, 9, 119
    triangle, 8, 119

B-spline, 140
    affine invariance, 150
    Bézier curve, 151, 152, 157
    basis function, 154
    biquadratic, 189
    de Boor algorithm, 143, 150, 157
    derivative, 151, 161
    differentiability, 150
    endpoint interpolation, 150
    evaluation, 143
    function, 153
    Greville abscissa, 153
    knot insertion, 156
    linear precision, 156
    local control, 143, 151
    local support, 156
    NURB, 195, 204
    partition of unity, 156
    periodic curve, 159
    properties, 150, 156
    rational, 204
    support, 153
    surface, 181

Bézier
    ordinate, 55
Bézier curve, 151
    cubic, 28
    de Casteljau algorithm, 32, 47, 83
    degree elevation, 50
    degree reduction, 52
    derivative, 30, 130
    evaluation, 32
    linear precision, 98
    properties, 28
    rational, 196
    representing a conic section, 196
    subdivision, 47
Bézier patch, 76
    2-stage de Casteljau evaluation method, 83
    2-stage explicit evaluation method, 76
    3-stage de Casteljau evaluation method, 84
    closed-form derivative, 80
    de Casteljau algorithm, 83
    degree elevation, 87
    degree reduction, 88
    properties, 77
    rational, 204
    ruled, 90
    tensor product, 78
Bézier patch
    tensor product, 99
Bézier polygon, 28
Bézier, Pierre, 27
barycentric combination, 4, 9, 210
barycentric coordinates, 9

basis function, 56
    B-spline, 140
    Bernstein, 28, 31
    Hermite, 69
    Lagrange, 63
    monomial, 39, 49, 62, 92
Bernstein polynomial, 28, 44, 55
Bessel tangent, 172, 187
bicubic Bézier patch, 76
bicubic Hermite interpolation, 103
bicubic patch, 96
bilinear interpolant, 78
bilinear interpolation, 101
bilinear precision, 78
binomial coefficients, 44
binormal vector, 116

Catmull-Clark algorithm, 190
Chaikin's algorithm, 188
Chinese character design, 134
chord length parameters, 68
circle, 199, 206
closed mesh, 211
complementary segment, 198
    of a circle, 200
cone, 208
conic section, 196
ConS, 105, 136
continuity
    C^1, 131, 179
    C^2, 132, 192
    G^1, 132
    G^2, 133
convert
    B-spline to Bézier, 157
    Bézier to monomial, 40
    Bézier to monomial, 49
    monomial to Bézier, 40, 49
convex hull, 20, 35
convex hull property, 29, 78, 151
Coons patch, 100
coordinate system, 1
corner cutting, 188
cross product, 6
cubic curve, 28

cubic Hermite polynomial, 69
curvature, 118, 133, 203
    Gaussian, 91
    mean, 124
    principal, 122
    signed, 119
curvature continuity, 133
curvature plot, 120
curve
    functional, 26, 54
    isoparametric, 76
    nonparametric, 55
    on surface, 75, 105, 136
    parametric domain, 26
cusp, 38

de Boor algorithm, 143, 150, 157,
    188
de Boor point, 140
de Boor, Carl, 143
de Casteljau algorithm, 32, 47, 83,
    188
de Casteljau, Paul de Faget, 32
debugging, 57, 209
decimation, 22
degree elevation, 50
    convergence, 52
    for surfaces, 87
degree reduction, 52
    for surfaces, 88
derivative
    Bézier curve, 130
    of B-spline curve, 161
    of Bézier curve, 33, 45
    of conic, 199
    of rational Bézier curve, 204
    of spline curve, 130
    partial, 79
developable surface, 91, 123
differentiability, 150, 152
digitizer, 20
domain, 14, 26, 71
domain knots, 141
Doo-Sabin algorithm, 189
dot product, 5

edge collapse, 22
ellipse, 198
elliptic point, 122
end conditions, 172, 174
    Bessel tangents, 172
endpoint interpolation, 28, 54, 150
equality check, 209
Euclidean space, 2
extraordinary vertex, 190
extrapolation, 29, 54

flatness test, 22
font, 135
forward difference, 31, 79, 80
Frenet frame, 116
functional curve, 26, 54, 56
functional surface, 72, 92

Gaussian curvature, 91, 122
generatrix, 205
global parameter, 16, 54, 129
graph
    of a function, 54
graph of a function, 26
Greville abscissa, 153

helix, 27
Hermite interpolation, 68, 135
    bicubic, 103
homogeneous coordinates, 196
hyperbola, 198, 219
hyperbolic paraboloid, 74
hyperbolic point, 122

IGES, 142
implicit form
    plane, 18
inflection point, 36, 119
internet, 23
interpolation, 59, 61, 63
    Hermite, 68
    bicubic, 95
    cardinal form, 63, 69
    endpoint, 28, 54, 78

    linear, 13, 14
interpolation constraints, 171
intersection, 106
isoparametric curve, 74, 76, 182
isophote, 126

junction point, 130

knot insertion, 156, 188
    Chaikin's algorithm, 188
knot multiplicity vector, 149
knot sequence, 130, 141
    domain knots, 141
    multiplicity, 141, 149
    simple knot, 141

Lagrange polynomial, 63
laser digitizer, 20
least squares approximation, 66, 106,
    166
length
    of vector, 6
light line, 125
Liming, R., 195
line
    explicit, 17
    implicit, 17
    parametric
      domain, 14
    parametric form, 14
    point–normal form, 17
    segment, 15, 20
linear combinations, 3
linear dependence, 6
linear interpolation, 13, 14, 65
linear precision, 29, 98, 156
linear space, 2
linear system, 96
    overdetermined, 67, 108, 167
linked list, 21
local control, 151
local parameter, 16, 54, 129
local support, 156
loop, 36

map
    affine, 7, 14, 29, 78
    linear, 8
    projective, 203
matrix form
    of Bézier curve, 48
mean curvature, 124
meridian, 206
midpoint, 4
minimal surface, 124
minmax box, 34
monomial form, 62
monomials, 39, 49
multiplicity, 141, 149, 150
multiresolution, 23

nonparametric curve, 55
nonparametric surface, 91
normal, 18, 135
normal curvature, 122
normal equations, 67, 109, 186
normal vector, 83, 116
normalized parameters, 68
numerical stability, 49
NURBS, 139, 195, 204

osculating circle, 118
osculating plane, 118

parabola, 196, 198, 201
parabolic point, 122
parallelogram rule, 11
parameter, 14, 141
    chord length, 68
    global, 16, 54, 129
    local, 16, 54, 129
    uniform, 68
parameter selection, 99, 111
parameter transformation, 16, 54
parameters
    normalized, 68
parametric form
    straight line, 16
partition of unitiy, 5
partition of unity, 57, 156

Pascal's triangle, 56
patch, 72
    Bézier, 75
    bicubic, 96
    bilinear, 73
    Coons, 100
    functional, 91
    Hermite, 103
periodic curve, 159
perpendicular, 6
perspective projection, 196
piecewise polynomial, 152
plane, 18
    implicit form, 18
    normal, 18
    point–normal form, 19
    point–vector form, 18
    tangent, 83
point, 2
    collinear, 7
    ratio of three, 7
polygon, 19
    B-spline, 140
    Bézier, 28
    closed, 20
    convex hull, 20, 35, 151
    open, 20
polynomial
    Bernstein, 44
    cubic Hermite , 69
    Lagrange, 63
    monomial, 62
potato chip, 122
preimage, 14
principal curvature, 122
projective invariance, 203

range, 14
ratio, 5, 7, 14, 131, 132
rational B'ezier curve
    derivative, 199
    reparametrization, 197
    representing a conic section, 198
    standard form, 197
rational B-spline curve, 204

rational B-spline surface, 204
  representing a surface of revo-
    lution, 205
rational Bézier curve, 201
  C^1 condition, 199
  curvature, 203
  torsion, 203
rational Bézier patch, 204
real space, 2
reflection line, 125
reparametrization, 197
ruled surface, 90

saddle point, 122
shape, 115
shape equations, 170, 186
smooth, 115
space
  affine, 2
  Euclidean, 2
  homogeneous, 203
  linear, 2
  real, 2
span, 141
spline curve, 130, 139, 195
  derivative, 130
standard form, 197
Stanford bunny, 13
star
  of a point, 22
STL, 22
straight line
  explicit form, 17
  implicit form, 17
  parametric form, 14
subdivision, 34, 47, 52
  Catmull-Clark algorithm, 190
  Chaiken's algorithm, 188
  Doo-Sabin algorithm, 189
  for surfaces, 88
subdivision curves, 188
subdivision surfaces, 188
  bicubic B-spline surface, 192
  biquadratic B-spline surface, 190
  extraordinary vertex, 190, 192

support, 153
surface
  developable, 91
  functional, 92
  nonparametric, 91
  of revolution, 205
  parametric
    domain, 71
  ruled, 90
  single valued, 92
  trimmed, 105
sweet, 115
symmetry, 28, 78

tangent plane, 83
tangent vector, 30, 78
tensor product, 78, 97, 99,
    181
tetrahedron, 10
tolerance, 84, 149
  for debugging, 210
torsion, 120, 203
torus, 206
translation, 3
translation invariance, 210
triangle, 8, 20
  area, 8
triangulation, 13, 20
  decimation, 22
  transmission, 23
trimmed surface, 105
twist vector, 81, 103

uniform parameters, 68
unit speed
  for rational curves, 200
unit square, 72
unit tangent, 116

vector, 2
vector product, 6

weight, 196

T - #0228 - 071024 - C0 - 235/191/13 - PB - 9780367455446 - Gloss Lamination